ATF Firearms Curios or Relics List

TABLE OF CONTENTS

COMPILER'S NOTES:

1. "*" indicates an entry made since ATF P 5300.11 was last published by way of hardcopy (10-95), supplemental, or Internet posting;
2. "S/N" and "S/Ns" mean "serial number" and "serial numbers;"
3. "Cal." and "cals." mean "caliber" and "calibers;"
4. "Mfd." means "manufactured."

This publication is a cumulative digest of determinations made by the Bureau of Alcohol, Tobacco, Firearms and Explosives (ATF) and is not inclusive of all weapons meeting curio or relic classification (i.e., firearms manufactured at least 50 years prior to the current date).

The "Introduction" section, under "Regulations Applicable to Licensed Collectors," refers to selected regulation citations applicable to licensed collector activities. The citations listed are illustrative and are meant to emphasize current regulations affecting licensed collectors. There is no intent to supercede law and regulation as described in The Gun Control Act of 1968 (GCA) as it is amended. Updated versions are published in ATF P 5300.4, Federal Firearms Regulations Reference Guide[1]. Questions arising from these citations should refer to this publication.

---

[1] ATF P 5300.4, Federal Firearms Regulations Reference Guide, can be obtained from the ATF Distribution Center (703) 455-7801.

INTRODUCTION

FEDERALLY LICENSED FIREARMS COLLECTORS

ATF is responsible for enforcing the provisions of the GCA and its subsequent amendments. A significant part of the GCA concerns the licensing and recordkeeping requirements pertaining to the manufacture, importation, distribution, and sale of firearms. A collector of Curios or Relics may obtain a collector's license under the GCA (see 18 U.S.C. Chapter 44 and the regulations issued thereunder in 27 CFR Part 478). The privileges conferred by this license extend only to transactions involving weapons classified as curio or relic firearms. In transactions involving firearms not classified as curios or relics, the licensed collector has the same status as a nonlicensee. A person need not be federally licensed to collect curios or relics. Generally, persons must be licensed in order to lawfully receive curios or relics from outside their State of residence.

The principal advantage of a Collector's License, therefore, is that a collector can acquire curios or relics in interstate commerce. Although a licensed collector may acquire and dispose of curios or relics at any location, dispositions to nonlicensees must generally be made to residents of the same State in which the collector is licensed. Further, **A LICENSED COLLECTOR IS NOT AUTHORIZED TO ENGAGE IN BUSINESS AS A DEALER IN ANY FIREARMS, INCLUDING CURIOS OR RELICS. A FEDERAL FIREARMS DEALER'S LICENSE IS REQUIRED FOR THIS ACTIVITY.** The term "engaged in business" as applied to a dealer in firearms refers, in part, to a person who devotes time, attention, and labor to engaging in such activity as a regular course of trade or business with the principal objective of livelihood and profit. Therefore, any person intending to "engage in the business" of selling firearms, including firearms defined by ATF as Curios or Relics, must first obtain a dealer's license. For example, if a collector acquires curios or relics for the purpose of sale rather than to enhance a collection, the collector would have to be licensed as a dealer in firearms under the GCA. **The sole intent and purpose of the collector's license is to enable a firearms collector to obtain curio or relic firearms from outside his or her State of residence.**

In order to obtain a collector's license, applicants need to prepare and submit ATF Form 7CR, Application for License (Collector of Curios or Relics), with the appropriate fee, $30 for 3 years, and in accordance with the instructions on the form. This form may be obtained from the Firearms and Explosives Licensing Center in Atlanta, Georgia, (404) 679-5040 or from your local ATF office.

WHAT ARE CURIOS OR RELICS?

As set out in the regulations (27 CFR 478.11), curios or relics include firearms which are of special interest to collectors by reason of some quality other than is associated with firearms intended for sporting use or as offensive or defensive weapons. To be recognized as curios or relics, firearms must fall within one of the following categories:

1.) Firearms which were manufactured at least 50 years prior to the current date, but not including replicas thereof;
2.) Firearms which are certified by the curator of a municipal, State, or Federal museum which exhibits firearms to be curios or relics of museum interest; and
3.) Any other firearms which derive a substantial part of their monetary value from the fact that they are novel, rare, bizarre, or because of their association with some historical figure, period, or event. Proof of qualification of a particular firearm under this category may be established by evidence of present value and evidence that like firearms are not available except as collector's items, or that the value of like firearms available in ordinary commercial channels is substantially less.

SELECTED REGULATIONS (27 C.F.R.) APPLICABLE TO LICENSED COLLECTORS

§ 478.11 Meaning of terms. Collector: Any person who acquires, holds, or disposes of firearms as curios or relics.

Collection premises: The premises described on the license of a collector as the location at which he maintains his collection of curios and relics. **Licensed collector:** A collector of curios and relics only and licensed under the provisions of this part. **Person:** Any individual, corporation, company, association, firm, partnership, society, or joint stock company.

§ 478.23 Right of entry and examination. (a) [A]ny ATF officer, when there is reasonable cause to believe a violation of the Act has occurred and that evidence of the violation may be found on the premises of any licensed manufacturer, licensed importer, licensed dealer, or licensed collector, may, upon demonstrating such cause before a Federal magistrate and obtaining from the magistrate a warrant authorizing entry, enter during business hours (or, in the case of a licensed collector, the hours of operation) the premises, including places of storage, of any such licensee for the purpose of inspecting or examining:

1) Any records or documents required to be kept by such licensee under this part and
2) Any inventory of firearms or ammunition kept or stored by any licensed manufacturer, licensed importer, or licensed dealer at such premises or any firearms curios or relics or ammunition kept or stored by any licensed collector at such premises.

(c) Any ATF officer, without having reasonable cause to believe a violation of the Act has occurred or that evidence of the violation may be found and without demonstrating such cause before a Federal magistrate or obtaining from the magistrate a warrant authorizing entry, may enter during hours of operation the premises, including places of storage, of any licensed collector for the purpose of inspecting or examining the records, documents, firearms, and ammunition referred to in paragraph (a) of this section (1) for ensuring compliance with the recordkeeping requirements of this part not more than once during any 12-month period or (2) when such inspection or examination may be required for determining the disposition of one or more particular firearms in the course of a bona fide criminal investigation. At the election of the licensed collector, the annual inspection permitted by this paragraph shall be performed at the ATF office responsible for conducting such inspection in the closest proximity to the collector's premises.

(d) The inspections and examinations provided by this section do not authorize an ATF officer to seize any records or documents other than those records or documents constituting material evidence of a violation of law. If an ATF officer seizes such records or documents, copies shall be provided the licensee within a reasonable time.

§ 478.25a Responses to requests for information. Each licensee shall respond immediately to, and in no event later than 24 hours after the receipt of, a request by an ATF officer at the National Tracing Center for information contained in the records required to be kept by this part for determining the disposition of one or more firearms in the course of a bona fide criminal investigation. The requested information shall be provided orally to the ATF officer within the 24-hour period. Verification of the identity and employment of National Tracing Center personnel requesting information may be established at the time the requested information is provided by telephoning the toll-free number 1-800-788-7133 or using the toll-free facsimile (FAX) number 1-800-578-7223.

§ 478.29 Out-of-State acquisition of firearms by nonlicensees. No person, other than a licensed importer, licensed manufacturer, licensed dealer, or licensed collector, shall transport into or receive in the State where the person resides (or if a corporation or other business entity, where it maintains a place of business) any firearm purchased or otherwise obtained by such person outside that State: **Provided,** That the provisions of this section:

(a) Shall not preclude any person who lawfully acquires a firearm by bequest or intestate succession in a State other than his State of residence from transporting the firearm into or receiving it in that State, if it is lawful for such person to purchase or possess such firearm in that State,
(b) Shall not apply to the transportation or receipt of a rifle or shotgun obtained from a licensed manufacturer, licensed importer, licensed dealer, or licensed collector in a State other than the transferee's State of residence in an over-the-counter transaction at the licensee's premises obtained in conformity with the provisions of § 478.96(c), and
(c) Shall not apply to the transportation or receipt of a firearm obtained in conformity with the provisions of §§ 478.30 and 478.97.

§ 478.29a Acquisition of firearms by nonresidents. No person, other than a licensed importer, licensed manufacturer, licensed dealer, or licensed collector, who does not reside in any State shall receive any firearms unless such receipt is for lawful sporting purposes.

§ 478.31 Delivery by common or contract carrier. (a) No person shall knowingly deliver or cause to be delivered to any common or contract carrier for transportation or shipment in interstate or foreign commerce to any person other than a licensed importer, licensed manufacturer, licensed dealer, or licensed collector, any package or other container in which there is any firearm or ammunition without written notice to the carrier that such firearm or ammunition is being transported or shipped:

Provided, That any passenger who owns or legally possesses a firearm or ammunition being transported aboard any common or contract carrier for movement with the passenger in interstate or foreign commerce may deliver said firearm or ammunition into the custody of the pilot, captain, conductor or operator of such common or contract carrier for the duration of that trip without violating any provision of this part.

(b) No common or contract carrier shall require or cause any label, tag, or other written notice to be placed on the outside of any package, luggage, or other container indicating that such package, luggage, or other container contains a firearm.

(c) No common or contract carrier shall transport or deliver in interstate or foreign commerce any firearm or ammunition with knowledge or reasonable cause to believe that the shipment, transportation, or receipt thereof would be in violation of any provision of this part:

Provided, however, That the provisions of this paragraph shall not apply in respect to the transportation of firearms or ammunition in in-bond shipment under Customs laws and regulation.

(d) No common or contract carrier shall knowingly deliver in interstate or foreign commerce any firearm without obtaining written acknowledgement of receipt from the recipient of the package or other container in which there is a firearm: **Provided,** That this paragraph shall not apply with respect to the return of a firearm to a passenger who places firearms in the carrier's custody for the duration of the trip.

§ 478.39a Reporting theft or loss of firearms. Each licensee shall report the theft or loss of a firearm from the licensee's inventory (including any firearm which has been transferred from the licensee's inventory to a personal collection and held as a personal firearm for at least 1 year), or from the collection of a licensed collector, within 48 hours after the theft or loss is discovered. Licensees shall report thefts or losses by telephoning 1-800-800-3855 (nationwide toll-free number) and by preparing ATF Form 3310.11, Federal Firearms Licensee Theft/Loss Report[2], in accordance with the instructions on the form. The original of the report shall be forwarded to the office specified thereon, and Copy 1 shall be retained by the licensee as part of the licensee's permanent records. Theft or loss of any firearm shall also be reported to the appropriate local authorities.

§ 478.41 General. (c) Each person seeking the privileges of a collector licensed under this part shall file an application, with the required fee (see § 478.42), with ATF in accordance with the instructions on the form (see § 478.44), and pursuant to § 478.47, receive from the Chief, National Licensing Center, the license covering the collection of curios and relics. A separate license may be obtained for each collection premises, and such license shall, subject to the provisions of the Act and other applicable provisions of law, entitle the licensee to transport, ship, receive, and acquire curios and relics in interstate or foreign commerce, and to make disposition of curios and relics in interstate or foreign commerce, to any other person licensed under provisions of this part, for the period stated on the license.

(d) The collector license provided by this part shall apply only to transactions related to a collector's activity in acquiring, holding, or disposing of curios and relics. A collector's license does not authorize the collector to engage in a business required to be licensed under the Act or this part. Therefore, if the acquisitions and dispositions of curios and relics by a collector bring the collector within the definition of a manufacturer, importer, or dealer under this part, he shall qualify as such.

§ 478.44 Original license. (b) Any person who desires to obtain a license as a collector under the Act and this part, or who has not timely submitted an application for renewal of the previous license issued under this part, shall file an application, ATF Form 7CR (Curios and Relics), with ATF in accordance with the instructions on the form. The application must be executed under the penalties of perjury and the penalties imposed by 18 U.S.C. 924. The application shall be accompanied by a completed ATF Form 5300.37 and ATF Form 5300.36 and shall include the appropriate fee

---

[2] ATF F 3310.11, Federal Firearms Licensee Theft/Loss Report, can be obtained from the ATF Distribution Center, (703) 455-7801.

in the form of a money order or check made payable to the Bureau of Alcohol, Tobacco, Firearms and Explosives. ATF Forms 7CR (Curios and Relics), ATF Forms 5300.37, and ATF Forms 5300.36 may be obtained by contacting any ATF office.

§ 478.47 Issuance of license. (a) Upon receipt of a properly executed application for a license on ATF Form 7, ATF Form 7CR, or ATF Form 8 Part II, the Chief, National Licensing Center, shall, upon finding through further inquiry or investigation, or otherwise, that the applicant is qualified, issue the appropriate license. Each license shall bear a serial number, and such number may be assigned to the licensee to whom issued for so long as the licensee maintains continuity of renewal in the same location (State).

(b) The Chief, National Licensing Center, shall approve a properly executed application for license on ATF Form 7, ATF Form 7CR, or ATF Form 8 Part II, if:

(1) The applicant is 21 years of age or over;

(2) The applicant (including, in the case of a corporation, partnership, or association, any individual possessing, directly or indirectly, the power to direct or cause the direction of the management and policies of the corporation, partnership, or association) is not prohibited under the provisions of the Act from shipping or transporting in interstate or foreign commerce, or possessing in or affecting commerce, any firearm or ammunition, or from receiving any firearm or ammunition which has been shipped or transported in interstate or foreign commerce;

(3) The applicant has not willfully violated any of the provisions of the Act or this part;

(4) The applicant has not willfully failed to disclose any material information required, or has not made any false statements as to any material fact, in connection with his application;

(5) The applicant has in a State

i.) premises from which he conducts business subject to license under the Act or from which he intends to conduct such business within a reasonable period of time, or in the case of a collector, premises from which he conducts his collecting subject to license under the Act or from which he intends to conduct such collecting within a reasonable period of time; and

(6) The applicant has filed an ATF Form 5300.37 (Certification of Compliance with State and Local Law) with ATF in accordance with the instructions on the form certifying under penalties of perjury that –

(i) The business to be conducted under the license is not prohibited by State or local law in the place where the licensed premises are located;

(ii) Within 30 days after the application is approved the business will comply with the requirements of State and local law applicable to the conduct of business;

(iii) The business will not be conducted under the license until the requirements of State and local law applicable to the business have been met; and

(iv) The applicant has completed and sent or delivered ATF F 5300.36 (Notification of Intent to Apply for a Federal Firearms License) to the chief law enforcement officer of the locality in which the premises are located, which indicates that the applicant intends to apply for a Federal firearms license. For purposes of this paragraph, the "chief law enforcement officer" is the chief of police, the sheriff, or an equivalent officer.

(c) The Chief, National Licensing Center, shall approve or the Director of Industry Operations shall deny an application for license within the 60-day period beginning on the date the properly executed application was received: **Provided,** That when an applicant for license renewal is a person who is, pursuant to the provisions of § 478.78, § 478.143, or § 478.144, conducting business or collecting activity under a previously issued license, action regarding the application will be held in abeyance pending the completion of the proceedings against applicant's existing license or license application, final determination of the applicant's criminal case, or final action by the Director on an application for relief submitted pursuant to § 478.144, as the case may be.

(d) When the Director of Industry Operations or the Chief, National Licensing Center, fails to act on an application for license within the 60-day period prescribed by paragraph (c) of this section, the applicant may file an action under section 1361 of title 28, United States Code, to compel ATF to act upon the application.

§ 478.49 Duration of license. The license entitles the person to whom issued to engage in the business or activity specified on the license, within the limitations of the Act and the regulations contained in this part, for a three year period, unless terminated sooner.

§ 478.50 Locations covered by license. The license covers the class of business or the activity specified in the license at the address specified therein. A separate license must be obtained for each location at which a firearms or ammunition business or activity requiring a license under this part is conducted except:

(a) No license is required to cover a separate warehouse used by the licensee solely for storage of firearms or ammunition if the records required by this part are maintained at the licensed premises served by such warehouse;

(b) A licensed collector may acquire curios and relics at any location, and dispose of curios or relics to any licensee or to other persons who are residents of the State where the collector's license is held and the disposition is made.[3]

(c) A licensee may conduct business at a gun show pursuant to the provision of § 478.100; or

(d) A licensed importer, manufacturer, or dealer may engage in the business of dealing in curio or relic firearms with another licensee at any location pursuant to the provision of § 478.100.

§ 478.58 State or other law. A license issued under this part confers no right or privilege to conduct business or activity contrary to State or other law. The holder of such a license is not by reason of the rights and privileges granted by that license immune from punishment for operating a firearm or ammunition business or activity in violation of the provisions of any State or other law. Similarly, compliance with the provisions of any State or other law affords no immunity under Federal law or regulations.

§ 478.91 Posting of License. Any license issued under this part shall be kept posted and kept available for inspection on the premises covered by the license.

§ 478.94 Sales or deliveries between licensees. A licensed importer, licensed manufacturer, or licensed dealer selling or otherwise disposing of firearms, and a licensed collector selling or otherwise disposing of curios or relics, to another licensee shall verify the identity and licensed status of the transferee prior to making the transaction. Verification shall be established by the transferee furnishing to the transferor a certified copy of the transferee's license and by such other means as the transferor deems necessary: **Provided,** That it shall not be required **(a)** for a transferee who has furnished a certified copy of its license to a transferor to again furnish such certified copy to that transferor during the term of the transferee's current license **(b)** for a licensee to furnish a certified copy of its license to another licensee if a firearm is being returned either directly or through another licensee to such licensee and **(c)** for licensees of multilicensed business organizations to furnish certified copies of their licenses to other licensed locations operated by such organization: **Provided further,** That a multilicensed business organization may furnish to a transferor, in lieu of a certified copy of each license, a list, certified to be true, correct, and complete, containing the name, address, license number, and the date of license expiration of each licensed location operated by such organization, and the transferor may sell or otherwise dispose of firearms as provided by this section to any licensee appearing on such list without requiring a certified copy of a license therefrom. A transferor licensee who has the certified information required by this section may sell or dispose of firearms to a licensee for not more than 45 days following the expiration date of the transferee's license.

§ 478.96 Out-of-State and mail order sales. (c)(1) A licensed importer, licensed manufacturer, or licensed dealer may sell or deliver a rifle or shotgun, and a licensed collector may sell or deliver a rifle or shotgun that is a curio or relic to a nonlicensed resident of a State other than the State in which the licensee's place of business is located if –

- (i) The purchaser meets with the licensee in person at the licensee's premises to accomplish the transfer, sale, and delivery of the rifle or shotgun;
- (ii) The licensed importer, licensed manufacturer, or licensed dealer complies with the provisions of § 478.102;

[3] Exception to this provision is provided for in **§ 178.96(c)(1)(i) in which a licensed collector is specifically authorized to sell a curio or relic shotgun or rifle to a nonlicensed resident of another State as long as the transaction is consummated at the licensed collector's premises and the transaction is legal in both States.**

(iii) The purchaser furnishes to the licensed importer, licensed manufacturer, or licensed dealer the firearms transaction record, Form 4473, required by § 478.124; and
(iv) The sale, delivery, and receipt of the rifle or shotgun fully comply with the legal conditions of sale in both such States.

§ 478.99 Certain prohibited sales or deliveries. (a) Interstate sales or deliveries. A licensed importer, licensed manufacturer, licensed dealer, or a licensed collector shall not sell or deliver any firearm to any person not licensed under this part and who the licensee knows or has reasonable cause to believe does not reside in (or if a corporation or other business entity does not maintain a place of business in) the State in which the licensee's place of business or activity is located. **Provided,** That the foregoing provisions of this paragraph **(1)** shall not apply to the sale or delivery of a rifle or shotgun (curio or relic, in the case of a licensed collector) to a resident of a State other than the State in which the licensee's place of business or collection premises is located if the requirements of § 478.96(c) are fully met, and **(2)** shall not apply to the loan or rental of a firearm to any person for temporary use for lawful sporting purposes. (See § 478.97)

(b) Sales or deliveries to underage persons. A licensed importer, licensed manufacturer, licensed dealer, or licensed collector shall not sell or deliver **(1)** any firearm or ammunition to any individual who the importer, manufacturer, dealer, or collector knows or has reasonable cause to believe is less than 18 years of age, and, if the firearm, or ammunition, is other than a shotgun or rifle, or ammunition for a shotgun or rifle, to any individual who the importer, manufacturer, dealer, or collector knows or has reasonable cause to believe is less than 21 years of age, or **(2)** any firearm to any person in any State where the purchase or possession by such person of such firearm would be in violation of any State law or any published ordinance applicable at the place of sale, delivery, or other disposition, unless the importer, manufacturer, dealer, or collector knows or has reasonable cause to believe that the purchase or possession would not be in violation of such State law or such published ordinance.

(c) Sales or deliveries to prohibited categories of persons. A licensed manufacturer, licensed importer, licensed dealer, or licensed collector shall not sell or otherwise dispose of any firearm or ammunition to any person knowing or having reasonable cause to believe that such person:

(1) Is, except as provided by § 478.143, under indictment for, or, except as provided by § 478.144, has been convicted in any court of a crime punishable by imprisonment for a term exceeding 1 year;
(2) Is a fugitive from justice;
(3) Is an unlawful user of or addicted to any controlled substance (as defined in section 102 of the Controlled Substance Act, 21 U. S. C. 802);
(4) Has been adjudicated as mentally defective or has been committed to any mental institution;
(5) Is an alien illegally or unlawfully in the United States; **Provided** That the provisions of this paragraph (c)(5) do not apply to any nonimmigrant alien if that alien is –
(i) Admitted to the United States for lawful hunting or sporting purpose or is in possession of a hunting license or permit lawfully issued in the United States;
(ii) An official representative of a foreign government who is either accredited to the United States Government or the Government's mission to an international organization having its headquarters in the United States or en route to or from another country to which that alien is accredited. This exception only applies if the firearm or ammunition is shipped, transported, possessed, or received in the representative's official capacity;
(iii) (iii) An official of a foreign government or a distinguished foreign visitor who has been so designated by the Department of State. This exception only applies if the firearm or ammunition is shipped, transported, possessed, or received in the official's or visitor's official capacity, except if the visitor is a private individual who does not have an official capacity; or
(iv) A foreign law enforcement officer of a friendly foreign government entering the United States on official law enforcement business;
(6) Has been discharged from the Armed Forces under dishonorable conditions;
(7) Who, having been a citizen of the United States, has renounced citizenship;
(8) Is subject to a court order that restrains such person from harassing, stalking, or threatening an intimate partner of such person or child of such intimate partner or person, or engaging in other conduct that would place an intimate partner in reasonable fear of bodily injury to the partner or child, except that this paragraph shall only apply to a court order that
(i) Was issued after a hearing of which such person received actual notice, and at which such person had the opportunity to participate; and

(ii) (A) Includes a finding that such person represents a credible threat to the physical safety of such intimate partner or child; or
(B) By its terms explicitly prohibits the use, attempted use, or threatened use of physical force against such intimate partner or child that would reasonably be expected to cause bodily injury,
or

(9) Has been convicted of a misdemeanor crime of domestic violence.

§ 478.103 Posting of signs and written notification to purchasers of handguns. (a) Each licensed importer, manufacturer, dealer, or collector who delivers a handgun to a nonlicensee shall provide such nonlicensee with written notification as described in paragraph (b) of this section.

(b) The written notification (ATF I 5300.2)[4] required by paragraph (a) of this section shall state as follows:

1.) The misuse of handguns is a leading contributor to juvenile violence and fatalities.
2.) Safely storing and securing firearms away from children will help prevent the unlawful possession of handguns by juveniles, stop accidents, and save lives.
3.) Federal law prohibits, except in certain limited circumstances,[5] anyone under 18 years of age from knowingly possessing a handgun, or any person from transferring a handgun to a person under 18.
4.) A knowing violation of the prohibition against selling, delivering, or otherwise transferring a handgun to a person under the age of 18 is, under certain circumstances, punishable by up to 10 years in prison.

§ 478.125 Record of receipt and disposition (f) Firearms receipt and disposition by licensed collectors. Each licensed collector shall enter into a record each receipt and disposition of firearms curios or relics. The record required by this paragraph shall be maintained in bound form under the format prescribed below. The purchase or other acquisition of a curio or relic shall, except as provided in [§ 27 C.F.R 478.125] (g) of this section, be recorded no later than the close of the next business day following the date of such purchase or other acquisition. The record shall show the date of receipt; the name and address or the name and license number of the person from whom received; the name of the manufacturer and importer (if any); the model, serial number, type, and the caliber or gauge of the firearm curio or relic. The sale or other disposition of a curio or relic shall be recorded by the licensed collector no later than 7 days following the date of such transaction. When such disposition is made to a licensee, the commercial record of the transaction shall be retained, until the transaction is recorded, separate from other commercial documents maintained by the licensee, and be readily available for inspection. The record shall show the date of the sale or other disposition of each firearm curio or relic, the name and address of the person to whom the firearm curio or relic is transferred, or the name and license number of the person to whom transferred if such person is a licensee, and the date of birth of the transferee if other than a licensee. In addition, the licensee shall:

1) Cause the transferee, if other than a licensee, to be identified in any manner customarily used in commercial transactions (e.g., a driver's license), and note on the record the method used, and
2) In the case of a transferee who is an alien legally in the United States and who is other than a licensee –
   (i) Verify the identity of the transferee by examining an identification document (as defined in § 478.11), and
   (ii) Cause the transferee to present documentation establishing that the transferee is a resident of the State (as defined in § 478.11) in which the licensee's business premises is located if the firearm curio or relic is other than a shotgun or rifle, and note on the record the documentation used or is a resident of any State and has resided in such State continuously for at least 90 days prior to the transfer of the firearm if the firearm curio or relic is a shotgun or rifle and shall note on the record the documentation used. Examples of acceptable documentation include utility bills or a lease agreement which show that the transferee has resided in the State continuously for at least 90 days prior to the transfer of the firearm curio or relic.
3) The format required for the record of receipt and disposition of firearms by collectors is as follows:

Firearms Collectors Acquisition and Disposition Record

Description of firearm							Disposition				
Manufacturer and/or importer	Model	Serial No.	Type	Caliber or gauge	Date	Name and address or name and license No.	Date	Name and address or name and license No.	Date of birth if nonlicensee	Driver's license No. or other identification if nonlicensee	For transfers to aliens, documentation used to establish residency

§ 478.126a Reporting multiple sales or other disposition of pistols and revolvers. Each licensee shall prepare a report of multiple sales or other disposition whenever the licensee sells or otherwise disposes of, at one time or during any 5 consecutive business days, two or more pistols, or revolvers, or any combination of pistols and revolvers totaling two or more, to an unlicensed person: **Provided**, That a report need not be made where pistols or revolvers, or any combination thereof, are returned to the same person from whom they were received. The report shall be prepared on Form 3310.4, Report of Multiple Sale or Other Disposition of Pistols and Revolvers[6]. Not later than the close of business on the day that the multiple sale or other disposition occurs, the licensee shall forward two copies of Form 3310.4 to the ATF office specified thereon and one copy to the State police or to the local law enforcement agency in which the sale or other disposition took place. Where the State or local law enforcement officials have notified the licensee that a particular official has been designated to receive Forms 3310.4, the licensee shall forward such forms to that designated official. The licensee shall retain one copy of Form 3310.4 and attach it to the firearms transaction record,[7] Form 4473, executed upon delivery of the pistols or revolvers.

§ 478.129 Record Retention (e) Records of dealers and collectors under the Act. The records prepared by licensed dealers and licensed collectors under the Act of the sale or other disposition of firearms and the corresponding record of receipt of such firearms shall be retained through December 15, 1988, after which records of transactions over 20 years of age may be discarded.

§ 478.147 Return of firearm. A person not otherwise prohibited by Federal, State, or local law may ship a firearm to a licensed importer, licensed manufacturer, or licensed dealer for any lawful purpose, and, notwithstanding any other provision of this part, the licensed manufacturer, licensed importer, or licensed dealer may return in interstate or foreign commerce to that person the firearm or a replacement firearm of the same kind and type. (See § 478.124(a) for requirements of a Form 4473 prior to return.) A person not otherwise prohibited by Federal, State, or local law may ship a firearm curio or relic to a licensed collector for any lawful purpose, and, notwithstanding any other provision of this part, the licensed collector may return in interstate or foreign commerce to the person the firearm curio or relic.

CURIOS OR RELICS AND COLLECTOR'S ITEMS: DEFINITIONS AND DETERMINATIONS

27 CFR § 479.24 <u>Destructive device determination</u>. The Director shall determine, in accordance with 26 U.S.C. 5845(f), whether a device is excluded from the definition of a destructive device. A person who desires to obtain a determination under that provision of law for any device which he believes is not likely to be used as a weapon shall submit a written request, in triplicate, for a ruling thereon to the Director. Each such request shall be executed under the penalties of perjury and contain a complete and accurate description of the device, the name and address of the manufacturer or importer thereof, the purpose of and use for which it is intended, and such photographs, diagrams, or drawings as may be necessary to enable the Director to make his determination. The Director may require the submission to him of a sample of such device for examination and evaluation. If the submission of such device is impracticable, the person requesting the ruling shall so advise the Director and designate the place where the device will be available for examination and evaluation.

27 CFR § 479.25 <u>Collector's items.</u> The Director shall determine, in accordance with 26 U.S.C. 5845(a), whether a firearm or device, which although originally designed as a weapon, is by reason of the date of its manufacture, value, design, and other characteristics primarily a collector's item and is not likely to be used as a weapon. A person who desires to obtain a determination under that provision of law shall follow the procedures prescribed in 27 CFR § 479.24 relating to destructive device determinations, and shall include information as to date of manufacture, value, design and other characteristics which would sustain a finding that the firearm or device is primarily a collector's item and is not likely to be used as a weapon.

27 CFR § 478.26 <u>Curio and relic determination.</u> Any person who desires to obtain a determination whether a particular firearm is a curio or relic shall submit a written request, in duplicate, for a ruling thereon to the Director. Each such request shall be executed under the penalties of perjury and shall contain a complete and accurate description of the

---

[6] ATF F 3310.4, Report of Multiple Sales or Other Disposition of Pistols and Revolvers, can be obtained from the ATF Distribution Center (703) 455-7801.

[7] Licensed collectors are not required to execute ATF Form 4473 for transfers of curios or relic firearms.

firearm, and such photographs, diagrams, or drawings as may be necessary to enable the Director to make a determination. The Director may require the submission of the firearm for examination and evaluation. If the submission of the firearm is impractical, the person requesting the determination shall so advise the Director and designate the place where the firearm will be available for examination and evaluation.

COLLECTORS WANTING A DETERMINATION

Collectors desiring to obtain a determination as to whether a particular firearm qualifies for classification as a curio or relic in accordance with 27 CFR 478.11 and 478.26 or a collector's item in accordance with § 479.25 and § 479.24 should submit a written request for a ruling. The letter should include:

1) A complete physical description of the item;
2) Reasons the collector believes the item merits the classification; and
3) Data concerning the history of the item, including production figures, if available, and market value.

As stated in the regulations, submission of the firearm <u>may</u> be required prior to a determination being made. Please submit your request to:

Bureau of Alcohol, Tobacco, Firearms and Explosives
Firearms Technology Branch
244 Needy Road
Martinsburg, WV 25405

IMPORTATION RESTRICTIONS

Licensed collectors may lawfully import curio or relic firearms other than surplus military firearms, nonsporting firearms, and NFA weapons. [A surplus military firearm is defined as one that belonged to a regular or irregular military force at any time. Alteration of the firearm does not change its status. Therefore, a sporting firearm with a surplus military frame or receiver is a surplus military firearm, because a frame or receiver is classified as a firearm as described in 18 U.S.C. § 921(a)(3).] Surplus military firearms are generally prohibited from importation under 18 U.S.C. § 925(d)(3). However, 18 U.S.C. § 925(e) authorizes licensed importers (FFL type 08 or 11) to import surplus military rifles, shotguns, and handguns classified as curios or relics; provided that such handguns are generally recognized as particularly suitable for or readily adaptable to sporting purposes. Nonsporting handguns are those pistols and revolvers that do not meet size and safety prerequisites, or that do not accrue a qualifying score on ATF Form 4590, "Factoring Criteria for Weapons." Surplus military firearms classified as curios or relics must be in their original military configuration to qualify for importation.

NFA weapons are those "firearms" defined in § 5845(a) of the National Firearms Act (26 U.S.C. Chapter 53). These include machineguns, silencers, shotguns having a barrel less than 18 inches in length, rifles having a barrel less than 16 inches in length, smooth bore handguns, disguised weapons, certain firearms having a bore greater than one-half inch in diameter, destructive devices (e.g., bombs, grenades, and mines), and certain other firearms. A licensed collector may not import NFA firearms.

Firearms may only be imported pursuant to a Form 6 issued by ATF. Collectors having questions concerning the importability of specific curio or relic firearms should contact:

Bureau of Alcohol, Tobacco, Firearms and Explosives
Firearms and Explosives Imports Branch
244 Needy Road
Martinsburg, WV 25405
Telephone: (304) 616-4550
Facsimile: (304) 616-4551

THE FIREARMS CURIOS OR RELICS LIST
THIS LIST IS COMPOSED OF FOUR MAJOR SECTIONS:

SECTION I. **Ammunition Classified as Curios or Relics:** Ammunition is no longer classified as "curios or relics" since the Congress in 1986 removed the interstate controls over ammunition under the GCA. We have continued the listing of 1972-1986 ammunition determinations as a service to those who rely on this list as a means of evaluation.

SECTION II. **Firearms Classified as Curios or Relics Under 18 U.S.C. Chapter 44:** Licensed collectors may acquire, hold, or dispose of these firearms as curios or relics. However, they are still "firearms" as defined in 18 U.S.C. 921(a)(3) and are, therefore, subject to all Chapter 44 controls. Generally, this category includes commemorative firearms, pistols, revolvers, rifles, and shotguns.

SECTION III. **Weapons Removed From the National Firearms Act as Collectors' Items and Classified as "Curios or Relics" Under 18 U.S.C. Chapter 44:** Weapons in this section are excluded entirely from the provisions of the National Firearms Act (NFA), 26 U.S.C. Chapter 53. Thus, approval from ATF to transfer these weapons is not required. They need not be registered in the National Firearms Registration and Transfer Record, and there is no transfer tax liability. However, these weapons are still "firearms" under Chapter 44, and remain subject to regulation under Part 478. Licensed collectors may receive these weapons in both intrastate and interstate commerce, and may transfer them intrastate, to both licensees and nonlicensees, and interstate, to licensed collectors and other licensees. While transfer may be made interstate to another licensee, they cannot be shipped interstate to a nonlicensee.

NOTE: Certain antique firearms are discussed and listed in Section IIIA.

SECTION IV. **National Firearms Act Weapons Classified as "Curios or Relics" Under 18 U.S.C. Chapter 44:** These weapons (e.g., machineguns) are firearms within the scope of the National Firearms Act (NFA), 26 U.S.C. Chapter 53, and are subject to all NFA Act provisions. Accordingly, these weapons cannot be lawfully transferred or received unless they are registered in the National Firearms Registration and Transfer Record at Bureau Headquarters. Once the registration requirements have been met, transfer may be made either intrastate or interstate to licensed collectors on ATF Form 4, "Application For Tax Paid Transfer and Registration of a Firearm." In each instance, first the appropriate transfer tax must be paid; ATF may then approve the transfer application. The application to transfer must be accompanied by an individual transferee's fingerprints and photograph as prescribed on the form. The form also requires a law enforcement official's certification, which must be completed.

Unserviceable weapons are still subject to the provisions of the National Firearms Act, except that they may be transferred free of transfer tax on ATF Form 5, "Application For Tax Exempt Transfer and Registration of a Firearm." As defined in 18 U.S.C. 5845(h) a firearm is unserviceable if it "is incapable of discharging a shot … [and] of being readily restored to a firing condition." Questions concerning the lawfulness of transactions in these weapons should be directed to:

Bureau of Alcohol, Tobacco, Firearms and Explosives
National Firearms Act Branch
244 Needy Road
Martinsburg, WV 25405
Telephone: (304) 616-4500
Facsimile: (304) 616-4501

Questions concerning the criteria for rendering a specific weapon unserviceable should be directed to:

Bureau of Alcohol, Tobacco, Firearms and Explosives
Firearms Technology Branch
244 Needy Road
Martinsburg, WV 25405
Telephone: (304) 616-4468
Facsimile: (304) 616-4443

NOTE: THE FOLLOWING LIST OF CURIOS OR RELICS IS NOT ALL-INCLUSIVE.

Questions concerning classification of a particular firearm should be directed to:

Bureau of Alcohol, Tobacco, Firearms and Explosives
Firearms Technology Branch
244 Needy Road
Martinsburg, WV 25405
Telephone: (304) 616-4468
Facsimile: (304) 616-4443

* * * * *

SECTION I: Ammunition Classified As Curios Or Relics

The following ammunition was classified as curios or relics prior to August 1986.

1. FOREIGN MILITARY RIFLE CARTRIDGES

.557 x 450 cal. Martini Henry
6.5 x 54Rmm Mannlicher (Dutch and Rumanian)
6.5 x 55mm Swedish Mauser
7.62 x 54Rmm Russian
7.92mm anti-tank cartridge for the German PzB 38, PzB 39, & Gr 39 anti-tank rifle.
7.92mm Kurz (7.92 x 33mm), as originally produced prior to 1946.
8 x 50Rmm Lebel
8 x 53mm Japanese Murata
8 x 56Rmm Portuguese Kropacheck, mfd. in Austria, for the Govt. of Portugal.
9.5 x 60mm Turkish Mauser
10.15 x 63mm Serbian Mauser
10.4 x 47Rmm Italian Vefterli M70
11 x 50mm Belgian Albini M67/72
11 x 53mm Belgian Comblain M71
11 x 60 Japanese Murata
11.15 x 60Rmm (.43) Mauser
11.3 x 50mm Beaumont M71
11.4 x 50mm Brazilian Comblain M74
11.43 x 55mm Turkish
11.7 x 51 mm Danish Remington
6.5 x 52mm Italian (Mannlicher-Carcano)
6.5 x 58mm Portuguese Vergueiro
7.5 x 54mm Swiss Schmidt-Rubin
7.65mm Argentine Navy Match 7.92mm anti-tank cartridge for the Polish (Marosczek) anti-tank rifle described as a cal. 7.92mm, Gr 39 Granatbuchse.
8 x 50Rmm Austrian Mannlicher
8 x 52Rmm Siamese Mauser, mfd. in Japan, for the Government of Siam
8 x 58Rmm Danish Krag
8 x 60mm Guedes M85 Portuguese
10.15 x 61mm Jarman
10.4 x 38Rmm Swiss Vefteril M69/81
10.75 x 58mm Russian Berdan
11 x 52mm Netherlands Beaumont M71/78
11 x 59mm French Gras
11.15 x 58mm Austrian Werndl M77
11.15 x 58mm (.43) Spanish Remington
11.4 x 50mm Austrian Werndl M73
11.43 x 50mm (.43) Egyptian Remington
11.5 x 57mm (.43) Spanish Reformado
12.11 x 44mm Remington M67 (Norway & Sweden)

2. EUROPEAN SPORTING RIFLE CARTRIDGES

6 x 29.5mm Stahl
6 x 58mm Forester (Rimless and Rimmed)
6.5 x 48mm Sauer
7 x 57 mm Mauser
6 x 57mm Mauser
6.5 x 27mm
6.5 x 53.5mm Daudeteau

3. BRITISH SPORTING RIFLE CARTRIDGES

.255 Rook Rifle
.369 Purdey Nitro Express
.375 Rimless Wesley Richards Nitro
.400/360 Purdey Flanged
.450 Nitro Express
.450 No. 2 Nitro Express
.476 Nitro Express
.600 Nitro Express
.360 No. 2 Nitro Express
.375/.303 Axite
.400/.360 Wesley Richards
.450 Black Powder Express
.450 Nitro for Black Powder Express
.475 No. 2 Jeffrey
.500 Black Powder Express

4. PISTOL AND REVOLVER CARTRIDGES

2.7mm Kolibri Auto
4.25mm Liliput Auto
5mm Clement Auto
6.5mm Bergmann, standard rimless 7mm Baby Nambu
.30/7.65mm Borchardt
.38 Long Colt
.44 Smith & Wesson (Martin Primer)
3mm Kolibri Auto
5mm Bergmann
5.5 Velo Dog Short
6.5mm Bergmann (without
extractor groove)7.65mm Roth-Sauer
.35 Smith & Wesson Auto
.44 Colt-Remington (Martin Primer)
.450 Long Revolver
.50 Remington (M71 Army)
9.8 Auto Colt
11mm German Service Revolver
11.75mm Montenegrin Revolver
.577 Pistol
11mm French Ordnance Revolver
11.25 Norwegian Raufoss (.45 ACP), pre-1946.

5. U.S. RIMFIRE CARTRIDGES

22 Rimfire cal. cartridges with experimental aluminum/steel cartridge cases.
.267 Rimfire cal. government experimental cartridge mfd. by Remington.
.32 Extra Short
.32 rimfire
.35 Allen
.38 Extra Long
.44 Short
.44 Long
.46 Long
.50 Peabody Musket
.50-70 Government
.54 Ballard
.56-50 Spencer
.56-52 Spencer, Tapered
.58 Joslyn Carbine
.58 Miller
.25 rimfire
.25 Short
.30 Long
.30 Shor.32 Long Rifle
.32 Extra Long
.38 rimfire
.41 Long
.44 Extra Long
.46 Extra Long
.46 Short
.50 Remington Navy
.52-70 Sharps
.56-46 Spencer
.56-52 Spener, Bottleneck
.56-56 Spencer
.58 Gatling Gun
.58 Mont Storm (Also known as .61 cal)

6. U.S. CENTERFIRE RIFLE CARTRIDGES

.22 Extra Long (Maynard)
.22-15-50 Stevens
.25-21 Stevens
.28 cal. centerfire experimental, mfd. at Frankford Arsenal.
.30-06 cal. cartridges mfd. by commercial manufacturers under government contract with headstamps indicating use for Palma Matches.
.30-30 Super X, Winchester "Bicentennial 76."
.30-30 Wesson
.308 Winchester, marked "Palma Match"
.32 Long Rifle
.32-30 Remington
.32-35 Maynard
.32-40 Winchester John Wayne .32-40 Bullard
.35 Winchester Self Loading

SECTION I

.38 Long
.38-35 Stevens
.38-50 Remington-Hepburn
.38-70 Winchester
.38-90 Winchester Express
.40-110 Winchester Express
.40-50 Sharps (Necked)
.405 Winchester
.40-60 Marlin
.40-60 Maynard 1882
.40-63 Ballard
.40-70 Sharps (Straight)
.40-70 Peabody "What Cheer"
.40-70 Remington
.40-72
.40-82
.40-90 Peabody "What Cheer"
.40-90 Sharps (Necked)
.40-90 Ballard
.44 Henry
.44 Extra Long Ballard
.44 Evans Long
.44-100 Wesson
.44-40 Extra Long
.44-60 Winchester
.44-70 Maynard
.44-77 Remington
.44-85 Wesson
.44-90 Remington (Stright 2.6" case)
.44-95 Peabody "What Cheer"
.45-50 Peabody (Sporting)
.40-65 Ballard Everlasting
.40-70 Sharps (Necked)
.22 Newton (Wildcat)
.25.36 Marlin
.25-25 Stevens
.28-30-120 Stevens
.30-03 Government (Note: Also known as Springfield)
.30-06 Commemorative, commemorating 50 years of the .30-06 cartridge as the standard U.S. service cartridge (headstamped USFA 1906-56).
.30-40 Wesson
.32 Ballard Extra Long
.32 Winchester Self Loading
.32-30 Maynard 1882
.32-35 Stevens
.32-40 Remington
.33 WCF
.35-40 Maynard
.38 Ballard Extra Long
.38-40 Remington-Hepburn
.38-45 Stevens
.38-50 Ballard
.38-56
.38-45 Bullard
.38-50 Maynard
.38-72
.40 Martin Primed
.40-40 Maynard 1882
.40-50 Sharps (Straight)
.70-150 Winchester
.40-70 Ballard
.40-70 Winchester
.40-75 Ballard
.40-85 Ballard
.40-90 Sharps (Straight)
.40-90 Ballard
.42 Martin Primer
.44 Wesson Extra Long
.44 Long C.F.
.44 Evans Short
.44-100 Ballard
.44-60 Sharps (Necked)
.44-60 Peabody "Creedmore"
.44-75 Ballard Everlasting
.44-77 Sharps
.44-90 Remington (Straight)
.44-90 Sharps (Necked)
.44-100 Remington "Creedmore" (Straight 2-6/10" case)
.45-75 Sharps (Straight)
Note: Identical to .45-70 Govt. Cartridge
.45-70 Van Choate
.45-78-475 Wolcoff
.45-80 Sharpshooter
.45-100 Sharps (Straight)
.45-120 Sharps (Stright 3-1/4" case)
.50 cal. experimental machine gun cartridge with aluminum cartridge case
.50 U.S. Carbine
.50-119 Sharps
.50-140 Winchester Express
.50-50 Maynard
.50-90 Sharps
.55-100 Maynard
.58 U.S. Musket (Berdan)
.45-100 Ballard
.45-100 Remington
.45-125 Winchester
.50 Government
.50 Remington
.50-100 Sharps
.50-115 Bullard
.50-140 Sharps
.50-70 Muske
.50-95 Winchester
.58 Berdan Carbine
.60 cal. experimental cartridge
6mm Lee Navy

SECTION I

7. MISCELLANEOUS CARTRIDGES

- Duplex cartridges in all calibers with the exception of 7.62mm NATO.
- All cartridges intended for use in "squeeze bore" firearms; e.g., Gerlach taper bore ammunition.
- Dardick cartridges (Trounds) in all calibers.

SECTION II: Firearms Classified As Curios Or Relics Under 18 U.S.C. Chapter 44

The Bureau has determined that the following firearms are curios or relics as defined in 27 CFR 47 8.11 because they fall within one of the categories specified in the regulations.

Such determination merely classifies the firearms as curios or relics and thereby authorizes licensed collectors to acquire, hold, or dispose of them as curios or relics subject to the provisions of 18 U.S.C. Chapter 44 and the regulations in 27 CFR Part 478. They are still "firearms" as defined in 18 U.S.C. Chapter 44.

Albanian SKS semiautomatic rifles, caliber 7.62 x 39, manufactured in Albania from 1964 to 1978.
Alkartasuna, semiautomatic pistol, caliber .32.
All Original military bolt action and semiautomatic rifles mfd. between 1899 and 1946.
All properly marked and identified semiautomatic pistols and revolvers used by, or mfd. for, any military organization prior to 1946.
All shotguns, properly marked and identified as mfd. for any military organization prior to 1946 and in their original military configuration only.
Argentine D.G.F.M. (FMAP) System Colt Model 1927 pistols, marked "Ejercito Argentino" bearing S/Ns less than 24501.
Argentine D.G.F.M. - (F.M.A.P.) System Colt model 1927, cal. 11.25mm commercial variations.
Armand Gevage, semiautomatic pistols, .32ACP cal. as mfd. in Belgium prior to World War II.
Astra, M 800 Condor model, pistol, caliber 9mm parabellum
Astra, model 1921 (400) semiautomatic pistols having slides marked Esperanzo Y Unceta.
Astra, model 400 pistol, German Army Contract, caliber 9mm Bergmann-Bayard, S/N range 97351-98850.
Astra, model 400 semiautomatic pistol, cal. 9mm Bergmann-Bayard, second German Army Contract, in S/N range 92851 through 97350.
Auto-Mag pistols, calibers .44 AMP and .357 AMP, mfd. and/or assembled by Autö-Mag Corporation, TDE, OMC,
High Standard, Lee Jurras, or AMT from 1969 to 1985.
Auto Ordnance, West Hurley, NY, Korean War Commemorative Thompson semiautomatic rifle, caliber .45.
Auto Ordnance, West Hurley, NY, World War II Commemorative Thompson semiautomatic rifle, caliber .45.
Auto Ordnance Thompson, cal. 45 semiautomatic rifle, Vietnam Commemorative, S/Ns V0001-V1500, issued by the American Historical Foundation, Richmond VA.
Baker Gun and Forging Company, all firearms mfd. from 1899 to 1919.
Bannerman model 1937, Springfield rifle, caliber .30-06.
Bayard, model 1923 semiautomatic pistol, cal. 7.65mm or .380, Belgian manufacture.
Beretta, M 1951 pistol Israeli Contract, caliber 9mm parabellum.
Beretta, model 1915 pistols, cal. 6.35mm, 7.65mm, and 9mm Glisenti.
Beretta, model 1915/1919 (1922) pistol (concealed hammer), caliber 7.65mm.
Beretta, model 1919 pistol (without grip safety), caliber 6.35mm.
Beretta, model 1923 pistol, caliber 9mm Glisenti.
Beretta, model 1932 pistol, having smooth wooden grips w/"PB" medallion, cal. 9mm.
Beretta, model 1934 pistol, light weight model marked "Tipo Alleggerita" or "All" having transverse ribbed barrel, cal. 9mm.
Beretta, model 1934 pistols, cal. 9mm post war variations bearing Italian Air Force eagle markings.
Beretta, model 1934 pistols, cal. 9mm produced during 1945 or earlier and having S/Ns within the ranges of 500000-999999, F00001-F120000, G0001-G80000, 00001AA-10000AA, 00001BB-10000BB. The classification does not include any post war variations dated subsequent to 1945 or bearing postwar Italian proof marks.
Beretta, model 1935 pistol, Finnish Home Guard Contract, marked "SKY" on the slide, cal. 7.65mm.
Beretta, model 1935 pistol, Rumanian Contract, marked "P. Beretta - cal. 9 Scurt - Mo. 1934 – Brevet." on the slide, cal. 9mm.

SECTION II

Beretta, Model 92SB with detachable folding shoulder stocks, cal. 9mm S/Ns C31509Z through C31538Z.

FN Browning, model 1902 (usually known as the model 1903) semiautomatic pistol, caliber 9mm Browning long.

Browning .22 caliber, semiautomatic rifles, Grade II, mfd. by Fabrique Nationale in Belgium from 1956 to 1976.

Browning FN Medalist Gold Line and "Renaissance" engraved semiautomatic pistols, caliber .22.

Browning FN Model 1910 semiautomatic pistols, calibers .32 and .380, marked "Browning Arms Company."

Browning Superposed Bi-Centennial Ltd. Edition, shotgun.

Browning, "Baby" model pistol, Russian Contract, caliber 6.35mm.

Browning, Centennial model High Power Pistol, cal. 9mm parabellum.

Browning, Centennial model 92 lever action rifle, cal. .44 Magnum.

Browning, model 1906 Pocket Pistol if more than 50 years old.

Browning, model 1922 pistol, cal. 7.65mm or 9mm Kurz, marked "C.P.I.M." denoting issue to the Belgian Political Police.

Browning, model 1922 pistol, cal. 7.65mm, bearing German NSDAP or RFV markings.

Browning, M1935 Hi Power pistol, Canadian, Congolese, Indian and Nationalist Chinese Contracts, cal. 9mm parabellum.

Browning Hi Power, Classic Edition, 9mm caliber pistol, having S/Ns 245BC0001 through 45BC5000.

Browning High Power D-Day Commemorative, 9mm caliber pistol, having S/Ns 245DD0001 through 245DD00150.

Browning High Power, Gold Classic Edition, 9mm caliber pistol, having S/Ns 245GC0001 through 245GC0500.

Browning, Superposed Centennial, consisting of a 20 gauge superposed shotgun, supplied with an extra set of .30-06 cal. superposed barrels.

Budischowsky, model TP70, semiautomatic pistol, cal. .25 ACP, with custom S/N DB1.

Campo-Giro, model 1913 and 1913/16 pistol, cal. 9mm Largo.

Chinese Communist, types 51 and 54 (Tokarev) pistols, cal. 7.62mm.

Chinese, Peoples Republic of China, copy of German Walther PPK .32 ACP cal. w/Chinese proof marks, Type I and II.

Chinese, Peoples Republic of China, copy of Japanese Type Sigiura Shiki semiautomatic pistol, caliber 7.65mm.

Chylewski, semiautomatic pistol mfd. by S.I.G. Switzerland, cal. 6.35mm (.25 ACP).

Clement, pistol, Belgian manufacture, cal. 5mm Clement.

Colt, Model 1911A1, cal. .45, semiautomatic pistols, Famous Generals of WWII, S/Ns WWIIG001 – WWIIG200.

Colt .38 National Match semiautomatic pistol, all S/Ns and in their original configuration.

Colt .38 Special Kit semiautomatic pistols, mfd. from 1964-1970, S/Ns 00100H-00434H.

Colt .45 ACP Kit semiautomatic pistols, mfd. from 1964-1970, S/Ns 001000-011640.

Colt, factory engraved for "Dr. Ramon Grau San Martin," President of Cuba, .45 cal. pistol with 5" barrel and blue steel finish; S/N C231769.

Colt, "Duke," Commemorative, .22 caliber revolver.

Colt, Ace semiautomatic pistol, cal. .22, mfd. by Colt from 1931 to 1947, S/N range from 1 to 10935 including those marked "UNITED STATES PROPERTY" on the right side of the frame.

Colt, Ace Service model semiautomatic pistol, cal. .22, mfd. by Colt from 1935 to 1945, S/N range from SM1 to SM13803 including those marked "UNITED STATES PROPERTY" on the right side of the frame.

Colt, Age of Flight 75th Anniversary semiautomatic pistols, cal. .45.

Colt, Aircrewman revolver produced between 1951 and 1959, cal. .38 Special, marked "Property of U.S. Air Force" on back strap, having Air Force issue numbers of 1 - 1189 and in the S/N range 1902LW – 90470LW.

Colt, Alabama Sesquicentennial, .22.

Colt, Alamo, .22 and .45.

Colt, American Combat Companion Officers model, cal. .45ACP pistol marked "1911 American Combat Companion 1981, 70 Years at America's Side."

SECTION II

Colt, American Combat Companion, Enlisted Man's model, cal. .45ACP pistol marked "1911 American Combat Companion 1981, 70 Years at America's Side."

Colt, American Combat Companion, General Officers model, cal. .45ACP pistol marked "1911 American Combat Companion 1981, 70 Years at America's Side." S/Ns 1 STAR, 2 STAR 3 STAR, 4 STAR, and 5 STAR.

Colt, Appomattox Court House Centennial, .22 and .45.

Colt AR-15 Sporter "The Viet Nam Tribute Colt Special Edition" .223 cal. semiautomatic rifle, bearing the American Historical Foundation registry numbers of VT0001 through VT1500.

Colt, Argentine model 1927, pistols, cal. .45, commercial variations.

Colt, Arizona Ranger model commemorative, .22 Revolver.

Colt, Arizona Territorial Centennial, .22 and .45.

Colt, Arkansas Territory Sesquicentennial, .22.

Colt, Army Model double action revolver, any cal., mfd. between 1899 and 1907.

Colt Army Special Model revolver, .32-20 cal., having a barrel length of 4'1/2", with factory engraving, S/N 329653.

Colt, ATF Special Edition, Deluxe model automatic pistol, cal. .45 ACP.

Colt, ATF Special Edition, Python Revolver, caliber .357 magnum.

Colt, ATF Special Edition, Standard model automatic pistol, cal. .45 ACP.

Colt, Abilene, .22 (Kansas City-Cow Town).

Colt, Bat Masterson, .22 and .45 (Lawman Series).

Colt, Battle of Gettysburg Centennial, .22.

Colt, Belleau Wood, .45 Pistol, (World War I Series).

Colt, Border Patrol model Revolver, .38 Special Heavy Duty, Police Positive (D) style frame, S/Ns within the range 610000 through 620000.

Colt, Buffalo Bill Historical Center, Winchester Museum, Special Issue, Colt Single Action revolver, cal. .44-40, S/N 21BB.

Colt, Buffalo Bill Wild West Show single action army .45 caliber.

Colt, California Bicentennial, .22.

Colt, California Gold Rush, .22 and .45.

Colt, Camp Perry Single Shot Target Pistols, .22 long rifle or .38 Special caliber.

Colt, Carolina Charter Tercentenary, .22 and .22/.45.

Colt, Chamizal Treaty, .22 and .45.

Colt, Chateau Thierry, .45 Pistol, (World War I Series).

Colt, Cherry's Sporting Goods 35th Anniversary, .22/.45.

Colt, Chisholm Trail, .22 (Kansas Series-Trails).

Colt, Civil War Centennial Single Shot, .22.

Colt, Coffeyville, .22 (Kansas Series-Cow Town).

Colt, Colorado Gold Rush, Colt, Colonel Samuel Colt, Sesquicentennial, .45.

Colt, Colt's 125th Anniversary, .45.

Colt, Columbus (Ohio) Sesquicentennial, .22.

Colt, Custom Gun Shop's "Custom Edition Sheriff's model" Single Action Revolver, cal. .45 Colt, S/Ns 1 to 35.

Colt, DA .38, New Army and Navy Revolver, made from 1899 to 1907.

Colt, Dakota Territory, .22.

Colt, Des Monies, Reconstruction of Old Fort, .22 and .45.

Colt, Detective Special revolver, cal. .38, S/N 418162, owned by Colonel Charles A. Lindbergh.

Colt, Detective Special revolvers, 2" barrels, marked "R.M.S. P.O. DEPT" or "U.S.P.O. DEPT."; S/Ns above 467000.

Colt, Dodge City, .22 (Kansas Series-Cow Town).

Colt, European Theater, .45 Pistol (World War II Series).

Colt, First model, Match Target Woodsman, cal. .22, semiautomatic pistol, mfd. from 1938 to 1944, S/Ns MT1 to MT15,000.

Colt, Florida Territory Sesquicentennial, .22.

Colt, Fort Findlay (Ohio) sesquicentennial, .22.

Colt, Fort Hays, .22 (Kansas Series-Forts).

Colt, Fort Larned, .22 (Kansas Series-Forts).

Colt, Fort McPherson (Nebraska) Centennial Derringer, .22.

SECTION II

Colt, Fort Scott, .22 (Kansas Series-Forts).

Colt, Fort Stephenson (Ohio) sesquicentennial, .22.

Colt, Fourth model Derringer, cal. .22 short rimfire, cased as a set of two pistols in a leather book titled "Colt Derringer, Limited Edition Colt, General George Meade, Pennsylvania, by Colt," on the spine of the book and "A Limited Edition by Colt," on the cover.

Colt, Forty-Niner Miner, .22.

Colt, General George Meade, Pennsylvania Campaign, .22 and .45.

Colt, General Hood, Tennessee Campaign Centennial, .22.

Colt, General John Hunt Morgan, Indiana Raid, .22.

Colt, General Nathan Bedford Forrest, .22.

Colt, Geneseo (Illinois) 125th Anniversary, Derringer, .22.

Colt, Golden Spike Centennial, .22.

Colt, Government model pistols in cal. .45 ACP, BB series.

Colt, H. Cook, "1 to 100," .22/.45.

Colt, Idaho Territorial Centennial, .22.

Colt, Indiana Sesquicentennial, .22.

Colt, J frame, Officers model Match, .38 Special revolver mfd. from 1970-1972, identified by J S/N prefix.

Colt, Joaquin Murrieta, "1 of 100," .22/.45.

Colt, John M. Browning Commemorative, .45 cal., semiautomatic pistol, S/Ns JMB 0001 – JMB 3000, and numbers GAS O JMB, PE CEW JMB, and 0003JMB.

Colt, John Wayne, Commemorative, .45 long Colt caliber revolver.

Colt, Kansas Centennial, .22.

Colt, Lightning model double action revolver, any cal. mfd. between 1899 and 1909.

Colt, Lightning rifles mfd. in 1899 through 1904.

Colt, Lord and Lady Derringer, .22 cal., as mfd. by Colts Patent Firearms Manufacturing Co., Hartford, CT.

Colt, Los Angeles Police Department (L.A.P.D.) Special Edition .45 cal. Government model semiautomatic pistol.

Colt, Maine Sesquicentennial, .22 and .45.

Colt, Mark IV, Government Model, commemorative "Michigan State Police 60th Anniversary, 1917-1977." The left side of slide engraved with scroll pattern and depicts 4 modes of transportation; Horse, Motorcycle, Auto and Helicopter S/Ns 1 to 1608.

Colt, Match Target Woodsman Semiautomatic Pistol, cal. .22LR., S/N 128866S, owned by Ernest Hemingway.

Colt, Meuse Argonne, .45 Pistol, (World War I Series).

Colt, Missouri Sesquicentennial single action army .45 caliber.

Colt, Missouri Sesquicentennial, .22.

Colt, Mk IV Series 70 semiautomatic pistols in all cals., which were incorrectly marked at the factory with both Colt Government model markings and Colt Commander markings.

Colt, model 1873 Peacemaker Centennial 1973, single action revolver, .44/.40 or .45.

Colt, model 1900 semiautomatic pistol, cal. .38, in original configuration.

Colt, model 1902 semiautomatic pistol, military model, cal. .38, in original configuration.

Colt, model 1902 semiautomatic pistol, sporting model, cal. .38, in original configuration.

Colt, model 1903 Pocket (exposed hammer), semiautomatic pistol cal. .38 ACP.

Colt, model 1903 Pocket (hammerless), semiautomatic pistol, cal. .32.

Colt, model 1908 Pocket (hammerless), semiautomatic pistol, cal. .380.

Colt, model 1908, cal. .25 ACP, hammerless semiautomatic pistol, having a grip safety, in S/N range 1 - 409061.

Colt, model 1911, commercial semiautomatic pistols, cal. .45 ACP, S/Ns Cl - C130000.

Colt, model 1911-A, commercial model, in cal. .45 and bearing Egyptian inscription meaning police, on the upper forward right-hand side of the trigger guard and having S/Ns within the range of C186000 to C188000.

Colt Model "Courier" double action revolver, .32 caliber, with 3-inch barrel.

Colt Model "Marshal" double action revolver, .38 special caliber, 2-inch and 4-inch barrels.

Colt, 50th Anniversary of the Battle of the Bulge, Model 1911A1, .45 caliber, semiautomatic pistol, having S/Ns between BB001 and BB300.

SECTION II

Colt, Montana Territory Centennial, .22 and .45.
Colt, National Match semiautomatic pistols, all S/Ns, in original configuration.
Colt, Nebraska Centennial, .22.
Colt, Ned Buntline Commemorative, cal. .45 revolver.
Colt, Nevada Centennial "Battle Born," .22 and .45.
Colt, Nevada Centennial, .22. and .45.
Colt, New Frontier .22 LR Revolvers, "Kit Carson" Commemorative, Colt model GB275.
Colt, New Frontier and Single Action Army model revolvers originally ordered & shipped with factory engraving, accompanied by a letter from the manufacturer confirming the authenticity of the engraving.
Colt, New Frontier, .357 magnum cal., single action revolver, barrel length 4-3/4" S/N 4411NF.
Colt, New Frontier, .45 cal., Abercrombie and Fitch, "Trailblazer."
Colt, New Jersey Tercentenary, .22 and .45.
Colt, New Mexico Golden Anniversary, .22.
Colt, New Police revolvers, .32 Colt cal., S/Ns 1 through 49,50.
Colt, New Service revolvers as mfd. between 1899 - 1944, all variations, all calibers.
Colt, NRA Centennial, Gold Cup National Match pistol, in cal. .45.
Colt, NRA Centennial, single action revolver, in cals. .357 Magnum and .45.
Colt, Officers model (1904-1930), .38 cal. revolver.
Colt, Officers model (1930-1949), .22 cal. revolver.
Colt, Officers model Match (1953-1969), .22 and .38 cal. revolvers.
Colt, Officers model Special (1949-1952), .22 and .38 cal. revolvers.
Colt, Officers model Target (1930-1949), .32 and .38 cal revolvers.
Colt, Officer's Model, .38 caliber revolver, S/Ns535472, 585683.
Colt Official Police Model revolver, cal. .32-20, having a barrel length of 5 inches, with factory engraving by Wilbur Glahn, S/N 554399.
Colt Official Police Model revolver, .38 Colt Special cal., having a barrel length of 6 inches, with factory engraving by Wilbur Glahn, S/N 554445.
Colt, Official Police Revolver, cal. .38, Silver Inlaid and Engraved by Wilbur A. Glahn, S/N 583469.
Colt, Oklahoma Territory Diamond Jubilee, .22.
Colt, Oregon Trail, .22 (Kansas Series-Trails).
Colt, Pacific Theater, .45 Pistol (World War II Series).
Colt, Pat Garrett, .22 and .45 (Lawman Series).
Colt, Pawnee Trail, .22 (Kansas Series-Trails).
Colt, Peacemaker Commemorative, .22 and .45 revolver.
Colt, Pocket Positive revolver, .32 cal.
Colt, Pocket Positive revolver, S/N 6164.
Colt, Police Positive .38 cal. revolvers, 2" barrels, marked "R.M.S. P.O. DEPT." or "U.S.P.O. DEPT."; S/Ns above 383000.
Colt, Police Positive revolver (1909); "St. L.P.D. #466" on butt; S/N 24466.
Colt, Pony Express Centennial, .22.
Colt, Pony Express, Russell, Majors and Waddell, Presentation model .45.
Colt, Python Model revolver, .357 cal., having a barrel length of 3 inches, with factory Type A engraving by Denise Thirion, in a presentation case with accessories, S/N T21895.
Colt, Python Revolver, cal. .357 Magnum, engraved and inlaid with the Crest of the United Arab Emirates.
Colt, Revolver, cal. .38, Police Positive, S/N 139212.
Colt, Rock Island Arsenal Centennial Single Shot, .22.
Colt, Santa Fe Trail, .22 (Kansas Series-Trails).
Colt model San Jacinto Special edition single action army revolver, .45 caliber, marked "THE BATTLE OF SAN JACINTO, APRIL 21, 1836" and "REMEMBER GOLIAD," 200 produced.
Colt, Second (2nd) Marne, .45 Pistol (World War I Series).
Colt, Shawnee Trail, .22 (Kansas Series-Trails).
Colt, Sheriffs model revolver, cal. .44 and .45.
Colt, Single Action Army (Bisley, Standard, and target variations), all original, mfd. from 1899 to 1946, S/N range from 182000 to 357869.

SECTION II

Colt, Single Action Army revolver, cal. .45, S/N 85163A, engraved and inlaid with a bust of President Abraham Lincoln.

Colt , Single Action Army (2nd Generation) revolvers, having S/Ns from 0001SA to 82000SA, all calibers, made between 1956 and 1976.

Colt, Single Action Revolvers, cal. .45, engraved & silver inlaid for presentation to Chuck Connors, S/Ns CC1 and CC2.

Colt, St. Augustine Quadricentennial, .22.

Colt, St. Louis Bicentennial, .22 and .45.

Colt Government model Texas Battleship Special Edition, .45 caliber pistols marked "BATTLESHIP EDITION - USS TEXAS," 500 produced.

Colt, Texas Ranger, .45.

Colt, Texas Sesquicentennial Standard and Premier model single action army, .45 cal.

Colt, the Liege Number 1 Colt Single Action Army Revolver, cal. .45, S/N Liege No. 1.

Colt, "The Right to Keep and Bear Arms" commemorative, .22 cal. Peacemaker Buntline, single action revolver having a 7-1/2" barrel with the inscription, "The Right to Keep and Bear Arms" inscribed on the barrel and a S/N range of G0001RB - G3000RB.

Colt, Theodore Roosevelt single action army .44-40 cal.

Colt, United States Bicentennial Commemorative, Python revolver, caliber .357.

Colt, United States Bicentennial Commemorative single action army revolver cal. .45.

Colt, West Virginia Centennial, .22 and .45.

Colt, Wichita, .22 (Kansas Series-Cow Town).

Colt, Wild Bill Hickok, .22 and .45 (Lawman Series).

Colt model Wisconsin State Patrol, 45th Anniversary, Special Edition Commemorative, 70 series, .45 government model pistols, consecutively numbered 1 through 126.

Colt, Woodman, cal. .22, semiautomatic target pistol, mfd. from 1915 to 1943, S/Ns 1 to 157000.

Colt, Woodsman, .22 caliber semiautomatic pistols, all models, all series, all S/Ns, (to include Match Target, Challenger, Huntsmen, and Targetsman), made prior to 1978.

Colt, Woodsman, First Model Match Target, .22 cal. semiautomatic pistols mfd. in or before 1944, a and having S/Ns MT1 through MT15100.

Colt, Wyatt Earp, .22 and .45 (Lawman Series).

Colt, Wyatt Earp, Buntline Special, .45 (Lawman Series).

Colt, Wyoming Diamond Jubilee, .22

Colt XIT Special Edition Single Action Army revolver, .45 caliber, barrel chemically etched with a covered wagon scene and "XIT RANCH," and marked "1 of 500."

Colt, 150th Anniversary single action army buntline, .45 cal.

Czechoslovakian, CZ27, 7.65mm semiautomatic pistol with Nazi markings.

Coston Supply Co., Coston Line Throwing Gun, .45-70 cal. S/N 447.

Czechoslovakian, CZ38, pistol cal. .380ACP.

Czechoslovakian, CZ50 pistol, cal. 7.65mm.

Czechoslovakian, CZ52 pistol, cal. 7.62mm.

Czechoslovakian, model 1952 and 1952/57, 7.62 x 45mm and 7.62 x 39mm cal., semiautomatic rifles.

Czechoslovakian, VZ-54 sniper rifle, manufactured from 1954 through 1957, caliber 7.62 x 54R, all S/Ns with scope mount and 2.5 power telescope.

Czechoslovakian, VZ82/CZ82 all calibers, all serial numbers.

* Daisy, Model V/L, .22 caliber caseless rifle, manufactured during 1968-1969.

Danish, M1910/1921 Bayard, pistol, cal. 9mm Bergmann-Bayard.

Davis Warner, Infallible, semiautomatic pistol, cal. .32.

Dreyse, semiautomatic pistols, all calibers.

Egyptian Raschid, semiautomatic rifle, cal. 7.62 x 39mm, original Egyptian military production.

Egyptian, Hakim (Ljungman) 7.92mm semiautomatic rifle as mfd. in Egypt.

Erma-Werke, Model EL 24, cal. .22, rifle, mfd. prior to 1946.

Esser-Barraft, English manufacture, slide action rifle, cal. .303.

Fabrique Nationale 1889-1989 Centenary High Power pistols, cal. 9mm.

Fabrique Nationale, Model 1906, .25 caliber semiautomatic pistol, all S/Ns.

Fabrique Nationale, model SAFN49 semiautomatic rifles, any caliber.

FN F.A.L. G and GL series, semiautomatic rifles, imported by Browning Arms Company, Arnold,

MO from 1959 to 1963, with the following S/Ns: G Series: G492, G493, G494, G537-G540, G649-G657, G662-G673, G677-G693, G709-G748, G752-G816, G848-G1017, G1021, G1033, G1035, G1041, G1042, G1174-G1293, G1415-G1524, G1570-G!784, G1800-Gl979, G1981-G1995, G2247-G2996, G3035-G3134. GL Series: GL749, GL835, GL1095-GL1098, GL1163,GL1164, GL1165, GL2004-GL2009, GL3135-GL3140.

Firepoint International, Ltd. Of England, 7.92x33 cal. S/Ns 08, 10, 11, 12, 14, 15, 16.

French Military Rifle Model 1949/56, in 7.62 x 51mm (NATO) cal. French, model 1949, cal. 7.5mm, semiauto. rifle (Fusil Mle. 1949 (MAS) 7.5mm).

French, model 1949/56 (Fusil Mle (MAS 7.5mm)) semiautomatic rifle.

A.H. Fox, Double barrel shotguns, all gauges, all grades, mfd. By Ansley H. Fox, Philadelphia, PA, and Savage Arms, Utica, NY, from approx. 1907 - 1947.

Geha and Rmo, shotguns made from Mauser rifles after World War I prior to 1946.

German military training rifles, cal. .22, single shot and repeaters, all manufacturers, in their original military configuration, marked "Kleinkaliber Wehersportsgewehr" (KKW), mfd. prior to 1946.

German sporting rifles, cal. .22, sporting rifles, single shot and repeaters, all manufacturers, in original configuration, marked "Deutsche Sportmodell" (DSM), prior to 1946.

German, model 1916 Grenatenwerfer original spigot type mortars.

German, P38 pistols, cal. 9mm parabellum mfd. prior to 1947.

Greener, Martini action, 14 gauge shotgun.

Gustloff, semiautomatic pistol in cal. 7.65mm mfd. by Gustloff Werke, Suhl, Germany.

Hammond or Grant Hammond, pistols, all models, variations or prototypes, made by Grant Hammond Corporation, New Haven, CT.

Hammond/Hi-Standard, semiautomatic pistols, in caliber .45.

Harrington and Richardson "Reising" Model 60 semiautomatic rifles, .45 ACP caliber, manufactured between 1944 and 1946.

Harrington and Richardson (H&R), Abilene Anniversary, .22 revolver.

Harrington and Richardson (H&R) Handy Gun, pistols with original rifled barrel, mfd. At Worcester, MA, all calibers, all barrel lengths, without shoulder stock.

Harrington and Richardson (H&R) Handy guns, mfd. at Worcester, MA, with shoulder stock, having an original smoothbore barrel 18 inches in length or greater, or original rifled barrel 16 inches in length or greater.

Harrington & Richardson Trapdoor Springfield carbine, .45-70 caliber, 100th Anniversary Little Big Horn Commemorative, manufactured between 1973-1981.

H&R, Centennial Officer's model Springfield rifle .45-70 Government.

H&R, Centennial Standard model Springfield rifle .45-70 Government.

H&R, model 999, revolver, cal. 22 Long rifle, barrel 6", 110th year commemorative, S/Ns from 001 to 999.

H&R, self loading semiautomatic pistol, caliber .32.

Hartford Arms and Equipment Company, single shot target pistol, caliber .22LR.

Hartford Arms and Equipment Company, repeating pistol, cal. .22LR.

Hartford Arms and Equipment Company, model 1928 pistol, cal. .22LR.

Hi-Standard, experimental electric free pistol, cal. .22 long rifle.

Hi-Standard, experimental electric free pistol, cal. .38 special.

Hi-Standard, experimental electric free pistol, cal. .38 special.

Hi-Standard, model P 38, semiautomatic pistol, cal. .38 SPL.

Hi-Standard, experimental model T-3 semiautomatic pistol, cal. 9mm Luger.

Hi-Standard, experimental ISU rapid fire semiautomatic pistol, caliber .22 short.

High Standard, Second Model Olympic pistol, cal. .22 short, mfd. between 1951 - 1953, S/Ns 330,000 – 439,999.

High Standard, Crusader Commemorative, Deluxe Pair, .44 Magnum and .45 Colt revolvers.

High Standard, model A pistol, caliber .22LR.

High Standard, model B pistol, caliber .22LR.

High Standard, model C pistol, caliber .22 Short.

High Standard, model D pistol, caliber .22LR.

High Standard, model E pistol, caliber .22LR.

High Standard, model H-A pistol, caliber .22LR.

High Standard, model H-B pistol, first model, caliber .22LR.
High Standard, model H-B pistol, second model, caliber .22LR.
High Standard, model H-D pistol, caliber .22LR.
High Standard, model H-E pistol, caliber .22LR.
High Standard, model USA-HD pistol, caliber .22LR.
High Standard, model HD-Military pistol, caliber .22LR.
High Standard, model G-380 pistol, caliber .380.
High Standard, model G-B pistol, caliber .22LR.
High Standard, model G-D pistol, caliber .22LR.
High Standard, model G-E pistol, caliber .22LR.
High Standard, model G-O (First model Olympic) pistol, caliber .22 Short.
High Standard, Supermatic Trophy model 107, .22 pistol Olympic Commemorative model.
High Standard, 1980 Olympic Commemorative, .22 caliber semiautomatic pistols, S/Ns USA1 - USA1000.
Holland and Holland, Royal Double Barrel Shotgun, .410 Gauge, S/N 36789.
Hopkins and Allen, model 1901, "FOREHAND," caliber .32 S&W long.
Hopkins and Allen, Revolver, .32 cal., S/N G 9545.
Hungarian Model 48 (Mosin Nagant M44 type) carbine, caliber 7.62 x 54R, manufactured in Hungary, and identified by the manufacturer code 02 on the chamber area and marked with the date of manufacture in the 1950's or earlier.
India rifle, cal. 7.62 2A, all variations, originally mfd. At Ishapore Arsenal, India prior to 1965.
Italian, Brixia, M1906, pistol, cal. 9mm Glisenti.
Italian, Glisenti, M1910, pistol, cal. 9mm Glisenti.
* Ithaca, Bicentennial Model 37, pump action shotguns, S/Ns USA0001 to USA1976.
Ithaca, double barrel shotguns actually mfd. in NY by the Ithaca Gun Co. Ithaca, NY. All gauges and all models, having barrels at least 18" in length and an overall length of at least 26", mfd. before 1950.
Ithaca Gun Co., single barrel trap guns, break open all gauges, all models actually mfd. at Ithaca, NY, before 1950.
Ithaca, St. Louis Bicentennial, model 49, .22 rifle.
Iver Johnson Arms, Pistol, cal. .380, U.S. Border Patrol 60th Anniversary commemorative, S/Ns USBP 0001 to USBP 5000.
Iver Johnson Arms, model M 1 Carbine, cal. .30, Korean War commemorative, S/Ns KW0001 to KW2500.
Iver Johnson Arms, model M 1 Carbine, cal. .30, Airborne commemorative, S/Ns KW0001 – KW2500.
Jieffeco, pistol, Belgian manufacture, caliber 7.65mm.
Jieffeco, semiautomatic pistol, cal. .25 ACP, marked "Davis Warner Arms Corp., N.Y."
Kimball, pistols, all models, all calibers.
Kolibri, pistols, cals. 2.7mm and 3mm Kolibri.
L. C. Smith, Shotguns mfd. by Hunter Arms Co. and Marlin Firearms Co. from 1899 to 1971.
Langenhan, semiautomatic pistols, all calibers.
Lahti, L-35 pistol, Finnish manufacture, caliber 9mm parabellum.
Lahti Swedish Model M40 pistols, cal. 9mm, all variations, mfd. prior to 1968.
Lee Enfield, No. 1 Mk III bolt action rifle, cal. .303, mfd. in Ishapore, India between 1946 and 1960. Original military configuration only.
Lee Enfield Rifle, caliber 7.62 2A and 2A1 "India" all variations, originally manufactured at Ishapore Arsenal, India, through 1973.
Lefever, shotguns made from 1899 to 1942.
Luger, Model 1902 Cartridge Counter, Mauser commercial, semiautomatic pistol, cal. 9mm, mfd. 1982.
Luger, pistol, all models and variations mfd. prior to 1946.
Luger, Mauser commercial manufacture, semiautomatic pistol, 70 Jahre, Parabellum-Pistole, Kelsoreich Russiand, commemorative, cal. 9mm.
Luger, Mauser commercial manufacture, semiautomatic pistol, 76 Jahre, Parabellum-Pistole, 1900-1975, commemorative, cal. 7.65mm.
Luger, Mauser commercial manufacture, semiautomatic pistol, 75 Jahre, Parabellum-Pistole,

SECTION II

Konigreich Bulgarian, commemorative, caliber 7.65mm.

Luger, Mauser Parabellum, semiautomatic pistol, 7.65mm or 9mm Luger, 4 and 6" barrel, Swiss pattern with grip safety and the American Eagle stamped on the receiver; made from 1970 to 1978.

MAB, model R pistol, caliber 9mm parabellum.

Makarov, pistol, Russian and East German, caliber 9mm Makarov.

Mannlicher, pistol, M1900, M1901, M1903 and M1905, caliber 7.63mm Mannlicher.

Marlin, Model 336 TS carbine, cal. 30-30, Powell Wyoming 75th anniversary commemorative, having PW S/N prefix.

Marlin, 90th Anniversary, model 39-A, .22 rifle.

Marlin, 90th Anniversary, model 39-A, .22 carbine.

Mauser, semiautomatic pistols mfd. prior to 1946, any caliber.

Mauser, Congolese model 1950 rifles marked FP 1952 on the receiver, caliber .30/06.

Mauser Luger, S/N 11.010034, 9mm, special engraving and ivory grips.

Mauser Luger, S/N RG 900/1001, 9mm, special engraving and walnut grips.

Mauser, model 1935 rifle 7 x 57mm cal. with Chilean Police Markings.

Mauser, P 38, pistols caliber 9mm, marked SVW46.

Mauser, rifles, bolt action and semiautomatic any caliber, commercially produced by Waffenfabrik Mauser, Oberndoff, Germany, prior to 1945.

MBA Gyrojet Carbine, S/N B5057.

MBA Gyrojet semiautomatic pistols, cal. 12mm or smaller, all models.

Menz, Liliput, German manufacture caliber 4.25mm.

Menz, PB III, in cal. 7.65mm, mfd. by August Menz, Suhl, Germany.

Menz, PB IIIA, in cal. 7.65mm, mfd. by August Menz, Suhl, Germany.

Menz, PB IV, in cal. 7.65mm, mfd. by August Menz, Suhl, Germany.

Menz, PB IVa, in cal. 7.65mm, mfd. by August Menz, Suhl, Germany.

Menz, Special, in cal. 7.65mm, mfd. by August Menz, Suhl, Germany.

Mexican, Obregon, pistol, caliber .45 ACP.

Mexican Model 1954 bolt action Mauser rifles and carbines, caliber .30-06 (original military configuration only).

Mossberg, Model 40, cal. .22 tubular fed, bolt action rifles mfd. from 1933-1935.

Mossberg, Model 42TR Targo, cal. .22 smoothbore, bolt action rifles, mfd. from 1940-1942.

Mossberg, Model 479RR "Roy Rogers" limited edition, cal. .30-30 lever action rifles, total number mfd. in 1983 only.

Mossberg, Model 25, .22 caliber bolt action rifles.

Mugica, model 120, pistol, caliber 9mm parabellum.

Navy Arms, Oklahoma Diamond Jubilee Commemorative, Yellow Boy Carbine.

Norinco (Chinese) AK47S, 5.56x45mm caliber, S/N 403876.

Norinco (Chinese) AK47S84S-1, 5.5 6x 45mm caliber, S/N 303052.

Norinco (Chinese) AK47S, 7.62 x 39mm caliber, S/N 1620127.

North Korean Type 1964, pistol, caliber 7.62mm Tokarev.

North Korean, Type 68, 7.62 x 25mm caliber semiautomatic pistols.

O.F. Mossberg Model 472SBAS "American Indian Commemorative" .30-30 and .35 Remington caliber lever action rifles, having Indian scenes etched on receiver, manufactured in 1974 only.

Ortgies, semiautomatic, caliber .25, with S/N 10073.

Ortgies, semiautomatic, caliber .32, with S/N 126314. OWA, semiautomatic pistol, caliber .25.

PAF, "Junior" semiautomatic pistol, caliber .25, mfd. by the Pretoria Arms Factory Ltd. of South Africa.

PAF, pistol, marked "BRF," caliber .25, mfd. by the Pretoria Arms Factory Ltd. of South Africa.

Parker-Hale, Model T-4, cal. 7.62mm, bolt action target rifles, mfd. prior to 1975.

Parker, shotguns, all grades, all gauges, produced by Parker Brothers, Meridan, CT, and Remington Arms, Ilion, NY, from 1899 through 1945.

Pedersoli, 120th Anniversary of the Remington Creedmore, .45/70 caliber, single shot rolling block rifle, having S/Ns between CR001 and CR300.

Pedersoli, 125th Anniversary of the Springfield, Model 1873 Trapdoor rifle, .45/70 caliber, having S/Ns between 125th- 001 through 125th-125.

SECTION II

Pedersoli, Sharps Creedmoor, single shot rifles, caliber .45-70, S/Ns SCR001 through SCR300.

* Pedersoli, Springfield Officer's Model trapdoor rifle, .45-70 caliber, having S/Ns 2NFM001 through 2NFM250.

Phoenix, (U.S.A.), pistol, caliber .25 ACP.Polish Mosin Nagant M44 type carbines, caliber 7.62 x 54R, manufactured in Poland, identified by the manufacturer code 11" in an oval on the chamber area, and marked with the actual date of manufacture during the 1950s or earlier.

Poly Tech (Chinese) AK 47S (386), 7.62 x 39mm caliber, S/N P47-11545.

James Purdey, Over & Under shotgun, 12 gauge, S/N 26819, engraved and gold inlaid.

Reising, .22 caliber, semiautomatic pistol.

Remington, No. 1, Mid Range Rolling Block target rifle reproduction, caliber .45-70, S/Ns beginnning with "RB97."

Remington, Model 31, pump action shotguns, 12, 16, and 20 gauge, mfd. 1931 - 1950.

Remington Rolling Block firearms, all models mfd. from 1899 - 1935.

Remington, over/under Derringer, caliber .41 rim fire, Remington Arms Company, Ilion, NY, made between 1898 and 1935.

Remington, Canadian Territorial Centennial, model 742, rifle.

Remington, model 12, rifle, cal. .22 short, long rifle, and .22 Remington Special mfd. by Remington Arms, Union Metallic Cartridge Co., Remington Works, Ilion, NY, from 1909 to 1936.

Remington, model 30 rifles.

Remington, model 720 rifles.

Remington, model 51, semiautomatic pistol, cals. .32 ACP or .380 ACP.

Remington, Montana Territorial Centennial, model 600, rifle.

Remington, 150th Anniversary model Nylon 66 semiautomatic rifle, caliber .22LR.

Remington, 150th Anniversary model 1100SA semiautomatic shotgun, caliber 12 gauge.

Remington, 150th Anniversary model 552A semiautomatic rifle, caliber. 22LR.

Remington, 150th Anniversary model 572A slide action rifle, caliber .22LR.

Remington, 150th Anniversary model 742ADL semiautomatic rifle caliber .30/06.

Remington, 150th Anniversary model 760ADL slide action rifle caliber .30/06.

Remington, 150th Anniversary model 870SA slide action shotgun, caliber 12 gauge.

Rheinmetal, semiautomatic pistols, caliber .32.

Rhode Island Arms Co., Morrone Model 46, shotgun.

* Romanian Model 56 semiautomatic carbine (SKS), caliber 7.62 x 39mm in original configuration as manufactured at Uzina Mechanica Cugir, Romania, from 1956 through 1962, having original

Romanian S/Ns consisting of two letters followed by up to four digits. The right side of the receiver is also marked with the year of production and the manufacturer's trademark consisting of an unfletched arrow within a triangle.

Roth Steyr, 1907, semiautomatic pistol, caliber 8mm.

Romanian AK 47S, 5.45 x 45mm caliber , S/N 3-040053-97.

Ruger, Blackhawk .44 with 6-1/2" barrel, revolver with S/Ns 1 to 29860.

Ruger, Blackhawk .44 with 7-1/2" barrel, revolver with S/Ns 17000 to 29860.

Ruger, Blackhawk .44 with 10" barrel, revolver with S/Ns 18000 to 29860.

Ruger, Blackhawk .357 with 4-5/8" barrel, revolver with S/Ns 1 to 42689.

Ruger, Blackhawk .357 with 6-1/2" barrel, revolver with S/Ns 20000 to 42689.

Ruger, Blackhawk .357 and 9mm stainless steel revolver with S/Ns 32-56000 - 32-59000.

Ruger, Blackhawk .357 magnum with 10" barrel, revolver with S/Ns 20000 - 38000.

Ruger, Canadian Centennial, Matched No. 1 Rifle Sets, Special Deluxe.

Ruger, Canadian Centennial, Matched No. 2 Rifle Sets.

Ruger, Canadian Centennial, Matched No. 3 Rifle Sets.

Ruger, Canadian Centennial, model 10/22, carbine.

Ruger, l0/22 Canadian Centennial, carbine with S/Ns Cl to C4500.

Ruger, Falling Block Long Range Creedmore rifle, cal. .45 (Sharps), S/N 130-06888, The Amber Silver Jubilee.

Ruger, flattop, "Blackhawk" revolvers, cals. .44 Magnum and .357 magnum, all barrel lengths, made from 1955 through 1962.

Ruger, flattop, single-six, .22 cal. revolvers with flat side loading gate, all barrel lengths, made from 1953 through 1956.

SECTION II

Ruger, Hawkeye, pistol with S/Ns 1 to 3296.
Ruger, Lightweight Single Six, Revolver with S/Ns 200000 to 212630.
Ruger, Mark I "U.S." stamped medallion, pistol with S/Ns 76000 - 79000.
Ruger, Single Six, engraved revolver with S/Ns 5100 - 75000.
Ruger, Standard Auto with red eagle, pistol with S/Ns 1 - 25000.
Ruger, Super Black Hawk, revolvers having a barrel length of 6½ inches with S/Ns 24000 - 26000.
Ruger, Super Single Six stainless with 4" barrel, revolver with S/Ns 62-07500 - 64-650000.
Ruger, Super Single Six stainless with 9" barrel, revolver with S/Ns 62-07500 - 63-40000.
Ruger, Super Single Six chrome with 4-5/8" barrel, revolver with S/Ns 504000 - 505000.
Ruger, "21 Club" No. 1, rifle with random S/Ns.
Ruger, 44 Deerstalker, carbine with S/Ns 1 to 5000.
Ruger, S47 Code, Model Super Blackhawk, 7 ½ inch barrel, .44 magnum revolvers, in S/N range 196 to 3111, with long grip frame, micro rear sight, and in mahogany wood case, approximately 500 manufactured.
Rumanian Mosin Nagant 1944 type carbines, caliber 7.62 x 54R, manufactured in Romania, and manufactured from 1952 to 1956.
Russian (U.S.S.R.), Nagant revolver, model 1895, cal. 7.62 Nagant and .22 cal. all variations, mfd. by the Tula Arsenal, Tula, Russia, after 1898.
Russian (U.S.S.R.), Tokarev, model TT, 1930, pistol, cal. 7.62, mfd. at the Tula Arsenal, Tula, U.S.S.R., from 1939 through 1956.
Russian (U.S.S.R.), Tokarev, model TT, 1933, pistol, cal. 7.62, mfd. at the Tula Arsenal, Tula, U.S.S.R., from 1933 through 1956.
Russian (U.S.S.R.) Tokarev, model TT R-3, .22 cal., pistol.
Russian (U.S.S.R.) Tokarev, model TT R-4, .22 cal., pistol.
Russian (U.S.S.R.), Tula Korovin, Tk, .25 ACP cal., semiautomatic pistol.
Russian (U.S.S.R.), model 1891, Mosin-Nagant rifles, cal. 7.62 x 54R and .22 cal., all models and all variations, mfd. after 1898 (i.e., M1891/30, M1910, M1938, and M1944).
Russian (U.S.S.R.), Tokarev, semiautomatic rifle, model 1938 (SVT38), cal. 7.62 x 54R, of Soviet manufacture.
Russian (U.S.S.R.), Tokarev, semiautomatic rifle, model 1940 (SVT40), cal. 7.62 x 54R, of Soviet manufacture.
Russian (U.S.S.R.), Tokarev, semiautomatic carbine, model 1932 (nonstandard), cal. 7.62 x 54R, of Soviet manufacture.
Russian (U.S.S.R.), Tokarev, semiautomatic carbine, model 1940 (SVT40), cal. 7.62 x 54R, of Soviet manufacture.
Russian (U.S.S.R.), Simonov, semiautomatic rifle, model SKS, cal. 7.62 x 39, of Soviet manufacture.
Russian (U.S.S.R.), Dragunov, semiautomatic rifle, model SVD, cal. 7.62 x 54R, of Soviet manufacture, Soviet military issue only.
* Sako, Anniversary Model, 7mm Remington magnum caliber, bolt action rifle.
J. P. Sauer & Sohn pistols, mfd. prior to 1946.
Sauer, 38(h), pistol, cal. 7.65mm marked w/Third Reich police acceptance stamps of Eagle C, F, K, or L. Savage Arms, semiautomatic pistols, cal. .45 ACP, all models.
Savage Arms, model 99, lever action, centerfire rifles, mfd. in Utica, NY prior to World War II with S/Ns below 450000.
Savage, Prototype pistols, cal. .25, .32 and .38 made between 1907 and 1927.
Savage model 1907 pistol, caliber .32 and .380.
Savage model 1915 pistol, caliber .32 and .380.
Savage model 1917 pistol, caliber .32 and .380.
Schwarzlose, pocket model 1908 in 7.65mm, pistol mfd. by A.W. Schwarzlose, G.m.b.h., Berlin, Germany, and those assembled or made by Warner Arms.
Smith and Wesson, Model 624 revolver, First Issue, cal. .44 Target, engraved year "1985" over the issue number (1-25) on the right sideplate.
Smith & Wesson Collector's Association, 25th Anniversary Commemorative revolver 1970-1995, Model 29, .44 Magnum caliber, S/Ns SWC 0001 through SWC 0184.
Smith & Wesson, 125th anniversary Commemorative, model 25, revolver, cal. .45, marked "Smith & Wesson 125th Anniversary" and mfd. in 1977.
Smith & Wesson, 150th anniversary Texas Ranger Commemorative model 19 revolver.

SECTION II

Smith & Wesson, 1st model, Ladysmith revolver, cal. .22 rimfire long.
Smith & Wesson, .22/32 Kit Gun, cal. .22LR, S/Ns 525670 - 534636 (no letter).
Smith & Wesson, 2nd model, Ladysmith revolver, cal. .22 rimfire long.
Smith & Wesson, 2nd model, single shot pistol, cals. .22 rimfire, .32 S&W and .38 S&W.
Smith & Wesson, .32 Double Action Top Break, cal. .32 S&W, S/Ns 209302 and higher.
Smith & Wesson, .32 Safety Hammerless Top Break (New Departure), cal. .32 S&W, S/Ns 91401 and higher.
Smith & Wesson, .357 Magnum Hand Ejector, cal. .357 Magnum, S/Ns 45768 to 60000 (no letter).
Smith & Wesson, .38 Double Action Top Break Perfected model, cal. .38 S&W.
Smith & Wesson, .38 Double Action Top Break, cal. .38 S&W, S/Ns 382023 and higher.
Smith & Wesson, .38 Hand Ejector Military and Police, cal. .38, S/Ns 1 to 241703 (no letter).
Smith & Wesson, .38 Safety Hammerless Top Break (New Departure), cal. .38 S&W, S/Ns 19901 and higher.
Smith & Wesson, .38/44 Outdoorsman & Heavy Duty, cal. .38, S/Ns 36500-62023 (no letter).
Smith & Wesson, 3rd model, Ladysmith revolver, cal. .22 rimfire long.
Smith & Wesson, 3rd model, single shot pistol, cals. .22 rimfire, .32 S&W and .38 S&W.
Smith & Wesson, 4 screw side plate revolvers, old style N-frame series, with no model designation stamped in the yoke cut, in cal. .44 magnum, all barrel lengths, falling within the S130000-S160350 block of S/Ns, of which a total of 6,500 units were produced from 1956 to 1958.
Smith & Wesson, .44 Hand Ejector, all cal., S/Ns 1-62488 (no letter).
Smith & Wesson, .455 Mark II Hand Ejector, caliber .455.
Smith & Wesson, California Highway Patrol Commemorative model 19 revolver, cal. .357.
Smith & Wesson, City of Los Angeles 200th Anniversary Commemorative model 19 revolver, cal. .357.
Smith & Wesson, K-22 Hand Ejector, cal. .22 LR, S/Ns 632132-696952 (no letter).
Smith & Wesson, K-32 Hand Ejector (K-32 Masterpiece), cal. .32 S&W Long, S/Ns 653388 to 682207 (no letter).
Smith & Wesson, Mercox Dart Gun, cal. .22 rimfire, blank.
Smith & Wesson, Model 16 (K-32 Masterpiece), cal. .32 S&W Long, "K" S/N series.
Smith & Wesson, model 21, .44 Special caliber, also known as the".44 Hand Ejector Fourth Model" and the "1950 Model Military," having S/Ns S75,000 - S263,000.
Smith & Wesson, Model .22/32 Hand Ejector (Bekeart model), cal. .22LR, S/Ns 138220 to 534636 (no letter).
Smith & Wesson, Model 29 "Elmer Keith Commemorative" .44 magnum revolvers, S/N EMK1 – EMK2500.
Smith & Wesson, Model 39, Connecticut State Police 75th Anniversary 1903-1978, pistols, S/Ns CSP001 – CSP704.
Smith & Wesson, Model 39, steel frame pistol, cal. 9mm parabellum.
Smith & Wesson, Model 39-1 (52-A), pistol, cal. 9mm parabellum.
Smith & Wesson, Model 53, Remington Jet Center Fire Magnum, cal. .22.
Smith & Wesson, Model 41-1 .22 short cal. semiautomatic pistols.
Smith & Wesson, Model 42 Centennial Airweight .38 Special 5-shot revolvers, with aluminum alloy frames and cylinders.
Smith & Wesson, Model 45 Military & Police .22LR cal. revolvers.
Smith & Wesson, Model 46 .22LR caliber semiautomatic pistols.
Smith & Wesson, Model 56 U.S. Air Force contract .38 Special 6-shot revolvers, S/N K500001 through K515001.
* Smith & Wesson Model 66 Bureau of Alcohol, Tobacco & Firearms 1933 – 1983 50th Anniversary Commemorative .357 magnum caliber revolvers, having a "BATF" prefix serial number and miniature replica of an agent's badge engraved into the frame.
Smith & Wesson Model 66 Distinguished Combat Magnum, caliber .357 magnum revolver, marked with the Texas Sheriff's Association badge and "TEXAS LAWMAN" on the right side of the frame, commemorating 150 years of law enforcement in Texas 1836-1986.
Smith & Wesson model 66, Dallas Police Department Commemorative Edition 1881-1981.
Smith & Wesson, Model 66, "Naval Investigative Service Commemorative" .357 6-shot revolvers.
Smith & Wesson, Model 147-A, 9mm 14-shot semiautomatic pistols.
Smith & Wesson, Model 544, "Texas Wagon Train Commemorative" .44/40 cal. 6-shot revolvers,

S/N TWT001 through TWT7800.

Smith & Wesson, .45 hand ejector model of 1950 Military, .45 caliber, having S/Ns between S76000 and S263000.

Smith & Wesson, Model 1917 revolver, cal. .45 ACP, produced for Brazil.

Smith & Wesson, Model Straight Line, single shot pistol, cal. .22 rimfire long rifle.

Smith & Wesson, pistol, caliber .32 ACP.

Smith & Wesson, pistol, caliber .35, all variations.

Smith & Wesson Registered Model 27 revolvers, cal. .357-magnum, 50 Yr. Commemorative, 5" barrel, S/Ns REG0001 through REG2500, inclusive.

Smith & Wesson U.S. Air Force contract M13 Aircrewman .38 Special 5-shot J frame and 6-shot K frame revolvers.

Smith & Wesson, U.S. Border Patrol 50th Anniversary Commemorative, model 66, stainless steel, cal. .357 Magnum, revolvers.

Sosso, pistols, mfd. by Guilio Sosso, Turin, Italy, or Fabrica Nationale D'Armi, Brescia, Italy, cal. 9mm.

Springfield Armory, Inc., Korean War Commemorative .30 cal., MI Garand Rifle S/Ns from KW0001 to KW1000.

Standard Arms Co., rifle/shotgun combination, U.S., model "Camp," slide action cal. .50.

Standard Arms Co., rifle model G, slide action or gas operated, any caliber.

Standard Arms Co., rifle model M, slide action caliber .25-.35, .30 Rem. and .35 Rem.

* Stevens, Models 425, 430, 435, and 440 High-Power Repeating Rifles in calibers .25 Remington, .30 Remington, .32 Remington, and .35 Remington.

Stevens, Model 77E, 12 gauge, military riot type shotguns, properly marked and identified as mfd. for the U.S. Military, from 1963 to 1969, in original configuration.

Steyr, model 1909, .25 ACP cal. semiautomatic pistol.

Steyr-Hahn, M1912, pistol, cal. 9mm Steyr.

Steyr-Hahn, M1912, pistol, cal. 9mm parabellum marked with Third Reich police acceptance stamps of Eagle C, F, K, or L.

Stock, semiautomatic pistols, all cals. mfd. by Franz Stock.

Swiss Model 1931/55 rifles, cal. 7.5 mm, S/Ns 1001 to 5150.

Swiss Schmidt Rubin, Model 1911, rifle made into a harpoon gun, caliber 12mm.

Swiss self loading rifle, test 1947, cal. 7.5mm, all variations.

Tauler, model military and police pistol.

Thompson/Center 25th Anniversary Contender, cal. .22LR pistol, 10 inch barrel, S/Ns 25 001-25 536.

Thompson Center Contender Pistol, cal. .30 Herrett, Steve Herrett Commemorative, S/Ns SH-001 to SH-500.

Thompson Center Contender Pistol, cal. 7mm TCU, IHMSA 10th anniversary commemorative, S/Ns IHMSA 10-001 to IHMSA 10-200.

Tippman Arms Company Models 1919 A-4, 1917, and .50 HB ½ scale, .22 caliber semi-automatic firearms, manufactured in Fort Wayne, Indiana, from 1986-1987.

Tokagypt 58, pistol, caliber 9mm parabellum.

Trejo, semiautomatic pistols, cal. .22, .32, and .380. mfd. in Mexico, circa 1952 to 1972.

* Uberti, 125th Anniversary of the Winchester, Model 1873 lever action rifle, .44/40 caliber, having S/Ns between A125th001 through A125th125.

Uberti Model 1866 lever action rifles, "An Engraver's Tribute to Gustave Young," .44/40 caliber,S/Ns GY001 through GY300.

Uberti model 1866 L.D. Nimschke, lever action rifles, caliber .44-40, S/Ns LDN001 through LDN300.

Uberti, Single action revolver, cal. .45, General George S. Patton Commemorative, S/Ns P0001 to P2500.

United States Patent Fire Arms Manufacturing Company, Artillery Model single action revolver, 100th anniversary of The Charge Up San Juan Hill, .45 caliber, with a 5 ½ - inch barrel and having S/Ns from NFM001 through NFM500 U.S, model 1911-A-1, .45 cal. pistol, mfd. by Union Switch and Signal Company, prototype model, with S/Ns US&S Exp. 1 to US&S Exp. 100.

U.S., model 1911-A1, semiautomatic, pistol, cal. .45, mfd. by the Singer Manufacturing Company in 1942, S/N range from S800001 to S800500.

SECTION II

U.S., model 1911-A1, semiautomatic pistol, cal. .45, mfd. by Remington Rand, bearing S/N prefix of ERRS.
U.S., model 1911-A1 semiautomatic pistol, cal. .45, produced as original factory cut-a-ways.
U.S., Rifle, cal. .30 M1, original military issue only, produced prior to 1956.
U.S. Rifle, caliber .30, M1, original issue only, produced prior to 1958.
U.S., Rifle, cal. .30, MC-1952, equipped with telescopic sight mount MC, telescopic sight MC1, marked U.S.M.C. or Kollmorgan.
U.S. Repeating Arms, Wyoming Centennial, Winchester Commemorative Model 94 Carbines, cal. .30-30, S/Ns WYC001 through WYC500.
UZI, Model A, semiautomatic carbine, cal. 9mm, having a satin nickel finish applied at the factory, S/Ns SA 0001 to SA 0100.
Walther, bolt action and semiautomatic rifles, all cals., mfd. prior to 1946.
Walther, model PP and PPK semiautomatic pistols, in all cals., mfd. in France and marked "MANURHIN".
Walther, Model PP pistol, 50 Jahre 1929-1959 Commemorative, caliber .22 and .380.
Walther, Model TP and TPH pistols, cal. .22 and .25 ACP, original German manufacture only.
Walther, Olympic bolt action single shot match rifle, in cal. .22 made by Waffenfabrik Walther, Zella-Mehlis (thur.) prior to World War II.
Walther, pistols, mfd. at Zel1a-Mehlis prior to 1946, all models, any caliber.
Walther, rifles, model 182, cal. .22 made by Waffenfabrik Walther, Zelia-Mehlis (thur.) prior to World War II.
Watson Brothers, Lee-Speed type, custom sporting rifle, cal. .303, S/N 7738.
Webley and Scott, model 1910 and 1913 high velocity pistols, cal. .38 ACP.
Webley and Scott, M1913, Navy or Commercial, self loading pistol, cal. .455.
Webley-Fosbury, semiautomatic revolvers, all cals., all models.
Webley, model 1909, pistol, cal. 9mm Browning Long.
Whitney, "Wolverine" and "Lighting" .22 cal. automatic pistols as mfd. by Whitney Firearms Company, Hartford, CT between 1955 - 1962.
Winchester, Model 1894, Florida Sesquicentennial carbine, cal. .30-30, S/Ns FL001-FL500.
Winchester, Apache Commemorative carbine, commemorative edition of model 1894 Winchester with S/N prefix of A.
Winchester, Comanche Commemorative carbine, commemorative edition of model 1894 Winchester with S/N prefix of CC.
Winchester, "Ducks Unlimited" shotgun, model 12, bearing S/Ns DU-001 through DU-800 (Commemorative).
Winchester, Kentucky Bicentennial Model 94 carbines, cal. .30-30, S/N range KY001-KY500.
Winchester, "Matched Set of 1000," a cased pair consisting of a Winchester model 94 rifle, cal. .30-30 and a Winchester model 9422 rifle, cal. .22.
Winchester, Model 12, pump action shotguns, mfd. from 1912 through 1963, S/Ns 1 through 1962017.
Winchester, Model 12, Shotgun, 12 gauge, prototype, for Ducks Unlimited Commemorative, S/N Y2002214.
* Winchester, Model 12 Y, pump action shotguns, all variations, manufactured between 1972 and 1979, S/Ns Y2000100 through Y2026399.
Winchester, Model 21, double barrel shotguns, all gauges, all grades, mfd. by Winchester and U.S. Repeating Arms.
Winchester, Model 21, Grand American Double Barrel Shotgun, cal. 20 and 28 gauge, S/N 32984, Engraved Custom Built by Winchester for Philip S. Rane.
Winchester, Model 37, single barrel shotguns, all gauges, manufactured between 1936 and 1963.
Winchester, Model 42, .410 gauge shotguns.
Winchester, Model 52, rifle, bearing S/Ns 1 to 6500.
Winchester, Model 53, all original, mfd. from 1924 to 1947 with 16" or longer barrel, and 26" or longer overall length.
Winchester, Model 54 rifles.
Winchester, Model 54, rifle, speed lock variation, cal. .270.
Winchester, Model 55, .22 caliber, single shot rifle.
Winchester, Model 61, cal. .22 rimfire, slide action repeater, hammerless.

Winchester, Model 62, cal. .22 rimfire, slide action repeater.
Winchester, Model 63, self loading rifles, cal. .22 rimfire.
Winchester, Model 64 and 65, lever action rifles.
* Winchester, Model 70, .300 Winchester Magnum, 50th Anniversary bolt action rifle, S/Ns 50 ANV 1 through 50 ANV 500.
Winchester, Model 70, bolt action rifle, all cals., mfd. in or before 1963 having S/Ns less than 581472.
Winchester, Model 70, rifle, cal. .308 rifle, 19" barrel and Mannlicher type stock, made from 1968 to 1971.
Winchester, Model 70, rifles, .308, .270 Winchester, and 30-06 cal., 19" barrel and Mannlicher type stock, made from 1968 to 1971.
Winchester, Model 70, Ultra Match Target Special Grade rifle, cal. .308.
* Winchester, Model 70, XTR Featherweight Ultra Grade, .270 caliber, bolt action rifle, 1 of 1,000.
Winchester, Model 71, all original, mfd. from 1936 to 1958 with 16" or longer barrel and 26" or longer overall length.
Winchester, Model 85, (single shot rifle), all original, mfd. from 1899 to 1920, with 16" or longer barrel and 26" or longer overall length.
Winchester, Model 86, all original, mfd. from 1899 to 1935, with 16" or longer barrel and 26" or longer overall length.
Winchester, Model 88, carbine, cal. .243, .284, .308, or .358 mfd. by Winchester Western Division, Olin Corporation, New Haven, CT.
Winchester, Model 88 rifles, all calibers.
Winchester, Model 92, all original, mfd. from 1899 to 1947, with 16" or longer barrel and 26" or longer overall length.
Winchester, Model 94, Alaskan Purchase Centennial, carbine.
Winchester, Model 94, American Bald Eagle Commemorative Carbine.
Winchester, Model 94, Arapaho Commemorative carbines, caliber .30-30, S/Ns ARAPA001 – ARAPA500.
Winchester, Model 94, Bat Masterson commemorative.
Winchester, Model 94, Bicentennial 76, carbine.
Winchester, Model 94, Buffalo Bill, carbine.
Winchester, Model 94, Buffalo Bill, rifle.
Winchester, Model 94, C.M. Russell, Great Western Artist Commemorative Carbine.
Winchester, Model 94, cal. .30-30, Antlered Game Commemorative, carbine.
Winchester, Model 94, cal. .30-30, Legendary Lawman Commemorative, carbine.
Winchester, Model 94, Calgary Stampede Commemorative, carbine, cal. .32 Winchester Special.
Winchester, Model 94, Canadian 1967, Centennial carbine.
Winchester, Model 94, Canadian 1967, Centennial rifle.
Winchester, Model 94, carbine, Canadian Pacific Centennial, cal. .32 Winchester Special.
Winchester, Model 94, carbine, Oklahoma Diamond Jubilee Commemorative.
Winchester, Model 94, carbines, Chevrolet Outdoorsman sets, .30-30 cal., having S/Ns between 5130000 and 5466000.
Winchester, Model 94, "Chief Crazy Horse," commemorative lever action rifle, cal. .38-55, mfd. by U.S. Repeating Arms Co., New Haven, CT.
Winchester, Model 94, Colt Commemorative Set, Winchester Signature model "Oliver F. Winchester" carbine, cal. 44-40, Lever Action, as mfd. by U.S. Repeating Arms Co., New Haven, CT.
Winchester, Model 94, Cowboy Commemorative, carbine.
Winchester, Model 94 rifle, Custom Limited Edition Centennial, .30 WCF cal., S/Ns CNTL 01 through CNTL 94.
Winchester, Model 94, Frederick Remington, Great Western Artist Commemorative Carbine.
Winchester, Model 94, Illinois Sesquicentennial, carbine.
Winchester, Model 94, in cal. .38/55, 1980, Alberta Diamond Jubilee Commemorative, carbine.
Winchester, Model 94, John Wayne Commemorative (Canadian Issue), carbine, cal. .32-40.
Winchester, Model 94, John Wayne commemorative, cal. .32-40, carbine.
Winchester, Model 94, Klondike Gold Rush Commemorative carbine.
Winchester, Model 94, Legendary Frontiersman rifle, cal. .38-55.
Winchester, Model 94, "Limited Edition I."
Winchester, Model 94, "Limited Edition II" rifle, cal. .30-30.

SECTION II

Winchester, Model 94, Limited Edition, carbine, cal. .30-30, S/Ns 77L1 - 77L1500.
Winchester, Model 94, Limited Edition Centennial, .30 WCF rifles, S/Ns CN10,001 - CN10,250, mfd. by the USRAC Custom Gun Shop for the Winchester Arms Collectors Association, having a replica Lyman No. 2 tang sight and with the WACA medallion placed in the buttstock.
Winchester, Model 94, Little Big Horn Centennial, carbine.
Winchester, Model 94, Lone Star Commemorative, carbine.
Winchester, Model 94, Lone Star Commemorative, rifle, .30-30.
Winchester, Model 94, model NRA Centennial, carbine.
Winchester, Model 94, Mounted Police, carbine.
Winchester, Model 94, Nebraska Centennial, carbine.
Winchester, Model 94, NRA Centennial rifle, .30-30.
Winchester, Model 94, One of One Thousand European Rifle commemorative.
Winchester, Model 94, Ontario, Canada Conservation Office, 100th Anniversary Commemorative carbines, caliber 30-30.
Winchester, Model 94, rifles and carbines mfd. prior to January 2, 1964, and having a S/N of less than 2,700,000, provided their barrel length is at least 16" and their overall length at least 26".
Winchester, Model 94, Royal Canadian Mounted Police Centennial carbine.
Winchester, Model 94, Saskatchewan Diamond Jubilee Carbine commemorative.
Winchester, Model 94, Sioux Commemorative carbines, cal. 30-30, with S/Ns beginning with "SU."
Winchester, Model 94, The Oliver F. Winchester commemorative.
Winchester, Model 94, Theodore Roosevelt, Carbine.
Winchester, Model 94, Theodore Roosevelt, Rifle.
Winchester, Model 94 "Trapper," .357 magnum caliber saddle ring carbine, 100th Anniversary Commemorative 1894-1994Winchester, Model 94, United States Border Patrol Commemorative carbine cal. .30-30.
Winchester, Model 94, Wild Bill Hickok Commemorative, Caliber .45, S/Ns WBH001 - WBH350.
Winchester, Model 94, Wells Fargo and Company Commemorative, carbines.
Winchester, Model 94, Wyoming Centennial Commemorative carbines, caliber .30-30, S/Ns WYC001 - WYC500.
Winchester, Model 94, Wyoming Diamond Jubilee, carbine.
Winchester, Model 94, Yellow Boy Indian, carbine.
Winchester, Model 94, 125th Anniversary Commemorative carbines, caliber .30-30, S/Ns WRAC001 - WRAC125.
Winchester, Model 94, 150th Anniversary Texas Ranger Commemorative, carbine.
Winchester, Model 95, all original, mfd. from 1899 to 1938, with 16" or longer barrel and 26" or longer overall length.
Winchester, Model 1866, Centennial, carbine.
Winchester, Model 1866, Centennial, rifle.
Winchester, Model 1873, all original, mfd. from 1899 to 1925, with 16" or longer barrel and 26" or longer overall length.
Winchester Model 1890 rifles.
Winchester, Model 1892 Carbine, .38 WCF cal. with 14" barrel S/N 654311.
Winchester, Model 1894, Golden Spike, carbine.
Winchester, Model 1894 Nez Perce Commemorative carbine, cal. .30-30, mfd. by U.S. Repeating Arms Co., S/Ns NEZ001-NEZ600.
Winchester, Model 1894, Texas Sesquicentennial rifle and carbine, cal. .38-55.
Winchester, Model 1894 Carbine, .30 WCF cal. with 15 1/16" barrel S/N 1040988.
Winchester, Model 1894 Special Carbine .30 WCF cal. S/N 948487.
Winchester, Model 1897, pump action shotguns, all gauges, manufactured between 1899 and 1957, S/Ns 1 through 1,024,700.
Winchester, Model 1897 or 97 riot guns, 12 gauge, 20" barrels with original RIC (Royal Irish Constabulary) markings.
* Winchester, Model 1902, .22 caliber bolt action rifles.
Winchester, Model 1903 .22 cal. semiautomatic rifles.
Winchester, Model 1906 rifles.
Winchester, Model 9422, Annie Oakley Commemorative Carbine .22 cal.

SECTION II

Winchester, Model 9422, Cheyenne Commemorative rifles, .22 cal.
Winchester, Model 9422, "Eagle Scout Commemorative" rifle, .22 cal.
Winchester, Northwest Territories Centennial rifle.
Yugoslavian M1948 (M48) bolt action rifles, 7.92 x 57mm caliber, produced at the Kragujevac Arsenal, original military configuration only.
Yugoslavian manufactured rifles M59 and M59/66, 7.62 x 39 mm caliber, all semiautomatic variations and having a fixed magazine, manufactured from 1947-1992.

SECTION III: Weapons Removed From The NFA As Collector's Items And Classified As Curios Or Relics Under The GCA

The Bureau has determined that by reason of the date of their manufacture, value, design and other characteristics, the following firearms are primarily collector's items and are not likely to be used as weapons and, therefore, are excluded from the provisions of the National Firearms Act.

Further, the Bureau has determined that such firearms are also curios or relics as defined in 27 CFR 478.11. Thus, licensed collectors may acquire, hold, or dispose to them as curios or relics subject to the provisions of 18 U.S.C. Chapter 44 and 27 CFR Part 478. They are still "firearms" as defined in 18 U.S.C. Chapter 44.

Any military bolt action or semiautomatic rifle mfd. prior to 1946 and accompanied by a "cup type" grenade launcher designed for the specific rifle.

Belgian, Pre-war mfd. Hi Power pistols, in cal. 9mm having tangent sights graduated to 500 meters, slotted for shoulder stock, having S/Ns of less than 47,000 without letter prefixes or suffixes and accompanied by original Belgian mfd. detachable wooden flat board type shoulder stocks.

Beretta, model 1918/1930, semiautomatic carbine, cal. 9mm, having a barrel length of 12.5" and a magazine capacity of 25 rounds.

Beretta, model 1923, semiautomatic pistol, in cal. 9mm Kurz (.380), accompanied by original Italian detachable leather and metal holster/shoulder stock.

Bergmann-Bayard, model 1908, Pistol, 9mm Bergmann-Bayard with shoulder stock and 4" barrel.

Bergmann, Mars model 1903, self loading pistol, w/accompanying shoulder stock.

British PIAT (Projector, Infantry, Anti-tank).

Browning, model 1903, Pistol, 9mm Browning Long, with shoulder stock and 5" barrel.

Browning Hi power pistols, 9mm having tangent sights graduated to 500 meters, slotted for shoulder stock, having S/Ns less than T200,000 etched vertically on the right side of slide, barrel, or frame and bearing crest of Emirates of Muscat & Oman, or mirror image of such crest, accompanied by original detachable wooden flat board shoulder stocks.

Canadian, Inglis No. 1, Chinese Contract, Hi Power pistols, cal. 9mm parabellum, having a tangent rear sight adjustable from 50 to 500 meters, slotted for shoulder stock, and having the letters C in the S/N and accompanied by original Canadian mfd. detachable wooden holster/shoulder stock.

Clement, Pistol Carbine, cal. 9mm.

Chinese, Jingal rifle, caliber .60, S/N IRS 9083H.

Chinese mfd. copies of the Mauser model 1896, semiautomatic pistol, produced prior to 1945, any cal., accompanied by original Chinese mfd. detachable wooden holster/shoulder stocks.

Colt, Lightning rifle, original cutaway demonstrator produced by Colt.

Colt, Model New Service, cal. .44 WCF, having a smooth bore, S/Ns 309065, 316627.

Colt, Model New Service revolver, cal. .45 Colt, S/N 326618, having a smooth bore.

Colt, Model New Service revolver, caliber .44 WCF, S/N 55125, having a barrel with a smooth bore.

Colt, Model Woodsman, .22 Long Rifle cal., semiautomatic pistol with an experimental 10" barrel and an experimental wooden detachable shoulder stock.

Colt, Model Woodsman, .22 Long Rifle cal., semiautomatic pistols, mfd. between 1915 and 1943, together with the original leather detachable holster stocks, mfd. by the N&S Corporation, Ventura, CA.

Colt, Model 1905, Pistol, .45 rimless, with leather holster/shoulder stock and 5" barrel.

Colt, Model 1911 pistol, cal. .38 Super, S/N 16212, accompanied by an original "Multi-Shot Riot and Anti-Bandit" shoulder stock as originally mfd. by Monarch Arms Sales Company, Los Angeles, California, in the early 1930s.

Colt, New Service Revolver, cal. 44/40, w/original smooth bore barrel, S/N 326596.

Colt, Officers model, .38 Special cal., double action revolver, with 6" barrel and a detachable, experimental skeleton shoulder stock and holster combination.

Colt, semiautomatic pistol, cal. .22 rimfire, S/Ns 49551 and 49712, accompanied by an original "Rifle-ette" shoulder stock as originally mfd. by Monarch Arms & Mfg Co., Los Angles,

CA in the 1920s.

Colt, Single Action Army revolver, cal. .45, with original smoothbore barrel, S/N 325085.

Colt, Single Action Army revolver, caliber 44 WCF (.44-40), S/N 324477, with blue finish, 5-1/2 inch barrel and smooth bore presented by Colt Firearms to Mr. John W. Garrett.

Colt, Single Action Army revolver, S/N 354096, cal. .44/40 having a smooth bored barrel and a barrel length of 7-1/2".

Czechoslovakian, model CZ24, semiautomatic pistol, in cal. 9mm Kurz (.380) accompanied by original Czechoslovakian detachable wooden holster/shoulder stock.

Fiala, model 1920, repeating pistol, cal. .22LR in all barrel lengths with accompanying detachable shoulder stock; original copies of the Fiala repeating pistol, marked Schall, Columbia or Botwinick Brothers, cal. .22LR, with accompanying original detachable shoulder stock.

Finnish, model L-35, Lahti, Semiautomatic pistol, in cal. 9mm parabellum, accompanied by original Finnish detachable wooden holster/shoulder stock.

Flashlight Revolver, cal. .22 Rimfire, having a rifled barrel, patented on July 10, 1923, and originally mfd. by S. P. Cottrell in Buffalo, NY.

Fromme, model 1912, semiautomatic pistol w/Benke-Thiemann folding shoulder stock.

German, model 1918, (WWI) anti-tank rifle (PzAgew 1918), cal. 13.25mm.

German, (Nazi) Belt Buckle Gun, .22 rimfire, marked "DRP Ausl Pat, Louis Marquis, W. Elberfeld."

German, (Nazi) Belt Buckle Gun, 7.65mm, marked "D.R.P. Angem."

German, Kamphpistole, cal. 26.7mm.

German, Leuchpistole, 26.7mm, Walther pattern, mfd. in or before 1945, with original 23mm rifle grenade launching adapter sleeve.

German, VG1-5 (Volksgewehr), semiautomatic rifle, cal. 7.92mm Kurz, having a barrel length of 14.9" and an overall length of 34.8".

Greener Cattle Killer (Original model) No. B1201, .310 cal.

Greener Cattle Killer (Pocket Pattern) No. B1203, .310 cal.

Greener Safti Killer No. B1216, .22 cal.

Greener Universal Safti Killer No. B1217, .310 cal.

Hamilton, models 7, 11, 15, 19, 23, 27, 027, 31, 35, 39, and 43 rifles.

Harrington and Richardson (H&R) "Handy Gun," .22 rimfire or .32-20 centerfire, with 12-1/4" rifled barrel, without shoulder stock.

Harrington and Richardson Handy Gun, Model 122 .22 Rimfire caliber, with original shoulder stock, having a barrel length of 12-3/4 inches, S/N 157.

Heal, Rifle No. 10, cal. .22 Heal, .22 rimfire cal. rifles, all models, mfd. prior to 1908, by the Heal Rifle Company or the Detroit Rifle Company.

High Standard, model C/S, smoothbore .22 cal. shot semiautomatic pistols, bearing S/Ns 59279, 59460, 59469, 59473, 59478.

High Standard, Model Sentinel (R-100), cal. .22 revolver having smooth bore with integral choke, S/Ns 948, 995.

High Standard, Model S, smoothbore .22 cal. shot semiautomatic pistols having slides marked "Hi-STANDARD MODEL "S" .22 L.R. SHOT ONLY," and bearing S/Ns 48142, 48143, 48144, 48145, 48146, 59474, 59496, 59458, 59459.

High Standard, Model S, smoothbore pistol, caliber .22 L.R., S/N 59493.

* Invisible Defender four barrelled palm gun in .22 caliber manufactured by O.F. Mossberg, Chicopee Falls, Massachusetts, circa 1902 – 1909.

Japanese, 20mm Single-Barrel Variant, flare pistol.

"JGA" (J.G. Anchutz, Ulm, Germany), .22 Flobert single shot pistol.

J.G. Anschutz, bolt action pistol, single shot, cal. 6 mm Glatt, S/N 377153.

Kennedy Repeating Rifle, cal. .44, with 14" barrel, S/N C329.

Krupp, models 1902 and 1906, 50mm Mountain Cannons as produced for the Siamese Government.

Luger, Artillery model, pistols having chamber dates of 1914 through 1918 or 1920, having German Weimar Navy markings consisting of the letter M over an anchor and a German Navy property number accompanied by original Artillery Luger flat board stocks, bearing German Weimar Navy markings of the letter M over an anchor with or without Navy property numbers.

Luger, the 1920 Commercial Artillery model, pistols as mfd. by DWM or Erfurt, having undated chambers, commercial proofmarks, and bearing the inscription Germany or Made in Germany on the receiver and accompanied by original, German mfd., artillery type, detachable wooden shoulder stocks.

Luger, DWM Pistol, model 1900, 1902, or 1906, in 7.65 Luger or 9mm parabellum cal., having the American Eagle chamber crest, and barrel lengths of either 4" or 4-3/4", with original detachable Ideal shoulder stocks and Ideal frame grips.

DWM Luger, Original models 1904, 1906, 1908, 1914, and 1920. Naval pistols in 9mm parabellum or 7.65mm cal., in both the Commercial and Naval military varieties; in both altered and unaltered barrel lengths in the model 1904 and in both altered and unaltered safety markings in the model 1906; with original board-type detachable shoulder stocks bearing brass or iron discs, with or without markings, or, if without brass or iron discs, being of the Navy flat board-type. This exemption applies only to the listed Naval Luger pistols if mated to the Naval Luger stock and will not apply if the Naval Luger pistol is mated to the Artillery stock. The Naval stock has an overall dimension of 12-3/4", a rear width of 4-5/8", a front width of 1-1/2", a rear thickness of 9/16", and a front thickness of 1-3/16".

Luger, DWM Stoeger model 1920 and 1923, semiautomatic pistols in 7.65mm or 9mm parabellum cal., in barrel lengths of 8, 10, 12, and 12-1/2", having either American Eagle chamber crests and/or Stoeger frame and/or upper receiver marks, having either standard, Navy or artillery rear sights, having extractors marked either "Loaded" or "Geladen" and having frame safety markings of either "Gesichert" or "Safe," together w/original commercial flat board stocks of the artillery type, which bear no S/Ns or military proof marks; may include a "Germany" marking.

Luger, DWM Pistol-Carbine, model 1920, 7.65mm or 9mm parabellum cal., with accompanying original commercial type shoulder stock, with or without forearm piece, having barrel lengths of 11-3/4" to less than 16".

Luger, German model 1914, Artillery model pistol, mfd. by DWM or Erfurt, having chambers dated 1914 –1918, bearing Imperial German military proofmarks & accompanied by original, German mfd., artillery type, detachable wooden shoulder stocks.

Luger, model 1902, Pistol-Carbine, 7.65mm Luger with original commercial type shoulder stock and forearm and 11-3/4" barrel.

Luger, Persian (Iranian) Artillery model, pistols, as mfd. by Mauser prior to 1945, accompanied by the original artillery type, detachable wooden shoulder stock, bearing a S/N in Farsi characters stamped into the wood on the left side.

Luger, semiautomatic pistol, certain variations with Benke-Thiemann folding shoulder stock.

MBA, Gyrojet Rocket Guns, cal. 13mm, semiautomatic version only, produced in 1968 or earlier, S/N ranges A0001-A0032, A001-A085, B010-B411, and B5059-B5692.

Manville, 18-shot-drum, 25mm, semiautomatic tear gas gun.

Marlin, Model 1892, caliber .44-.40, S/N 386485, with 15-inch barrel.

Marlin Model 1893 or 93 carbine, cal. .25, with 15 inch barrel, S/Ns 193134, 284957, 320283, 387186.

Marlin Model 1893 or 93 carbine, cal. .30, with 15 inch barrel, S/Ns 199669, 199690, 203990, 204585, 208526, 228856, 237681, 244372, 256559, 256568, 256577, 256583, 256611, 258853, 261844, 261889, 261901, 269397, 269636, 281802, 281810, 291358, 336467, 336478, 336504, 336610, 338883.

Marlin Model 1893 carbine, .30-30 caliber having a factory fitted 15-inch barrel S/N 1934.

Marlin Model 1893 carbine, cal. .30-30, with a 15 inch barrel, S/Ns 244371, C2231 and 269397.

Marlin Model 1893, carbine, cal. .32 with 15 inch barrel, S/N C2315.

Marlin Model 1893 or 93 carbine, cal. .32-30, .32-40, .32 HPS, with 15 inch barrel, S/Ns *2293, 176656, 194471, 194786, 194771, 194786, *251864, 258880, 258894, 285305, 285331, 291337, 291399, 292390, 307905, 308094, 308111, 329063, 335434, 336651, 336964, 370263, 405878, 426311, C1326, C1323, C3967.

Marlin Model 1893 or 93 carbine, cal. .38-55 or .38 HPS, w/15 inch barrel, S/Ns 194783, 246363, 75896, 316309, 336529, A4054, C3394, D8266.

Marlin Model 1894 or 94 carbine, cal. .25, with 12 inch barrel, S/Ns 283149, 328901, 328925, 328928, 328937, 328958.

Marlin Model 1894 or 94 carbine, cal. .25, with 15 inch barrel, S/Ns 190958, 190962, 190976, 211087, 280125, 280128, 280141, 280147, 283330, 287748, 294716, 328910, 337955, 442611, 442643.

SECTION III

Marlin Model 1894, .25-20 caliber, S/Ns 409198, 442606, having a 15-inch barrel.

Marlin Model 1894 or 94 carbine, cal. .32, with 15 inch barrel, S/Ns 190944, 190950, 190963, 190972, 217948, 282898, 282904, 287741, 287745, 287753, 287755, 287762, 287764, 287769, 287770, 287772, 294666, 294732, 301560, 314652, 325609, 328807, 328837, 328933, 376900.

Marlin Model 1894, caliber .32-20, S/N 435220, with 15-inch barrel.

Marlin Model 1894 carbine, caliber .33-20, S/Ns *D8959 and 417837, having a 15-inch barrel.

Marlin Model 1894 or 94 carbine, cal. .38, with 14 inch barrel, S/Ns 324849, 326718.

Marlin Model 1894 or 94 carbine, cal. .38, with 15 inch barrel, S/Ns 195570, 199975, 200007, 20022, 215011, 216362, 216412, 282004, 282936, 282997, 285870, 285974, 285979, 285990, 286281, 287729, 300626, 310864, 316798, 318000, 321957, 321970, 321973, 321985, 321986, 321990, 322025, 322035, 322037, 322053, 322077, 322079, 322082, 322095, 322514, 323672, 323688, 323785, 323810, 323846, 323925, 323928, 324195, 324238, 325732, 325740, 326143, 326206, 326726, 330543, 384186, 397215.

Marlin Model 1894 or 94, caliber .38-40, S/N 430591, as originally manufactured with a 15-1/4 inch barrel.

Marlin Model 1894 or 94 carbine, cal. 44, with 14 inch barrel, S/Ns 312691, 312708, 312801, 316617, 316628, 316632, 316645, 316649, 316654, 316656, 316739, 316772, 316803, 317427, 317449, 317466, 317470, 317499, 317559, 317573, 317578, 317582, 317587, 317596, 317646, 317687, 317690, 317698, 317697, 322641.

Marlin Model 1894 or 94 carbine, cal. .44, with 15 inch barrel, S/Ns 181852, 181863, 181911, 181967, 182048, 182068, 182093, 182205, 182219, 182226, 182321, 182322, 182344, 183374, 183376, 183379, 183400, 183401, 183403, 183408, 183413, 183418, 183428, 183432, 183433, 185128, 185129, 185133, 185145, 185150, 185155, 185165, 185167, 185174, 185175, 185185, 85186, 185187, 185192, 191033, 191096, 191097, 191098, 191099, 191100, 191101, 191104, 191105, 191106, 191424, 195823, 195840, 195842, 196827, 197867, 200830, 200903, 200906, 200922, 200928, 200929, 200930, 200958, 203760, 203761,203763, 203766, 203767, 203769, 203771, 203772, 203773, 203774, 203778, 203780, 203781, 203782, 203784, 203785, 203787, 203788, 203789, 203796, 203798, 203806, 203811, 203812, 203813, 203818, 203822, 203824, 203826, 204395, 204398, 204403, 204404, 204407, 204411, 204414, 204415, 204419, 204421, 204426, 204427, 204428, 204429, 204430, 204499, 205727, 210260, 210269, 211064, 215017, 215019, 215020, 215023, 215030, 215037, 215043, 215054, 215058, 215059, 215063, 215068, 215070, 215071, 215079, 216340, 216348, 216350, 216357, 216363, 216366, 216368, 216416, 216423, 216424, 216425, 219240, 219253, 219254, 219273, 219288, 219291, 219295, 232678, 232948, 232950, 232964, 232981, 232997, 232999, 233001, 233010, 233240, 233281, 233299, 233344, 233355, 233369, 233396, 233403, 233481, 233519, 233530, 233535, 233543, 233549, 233592, 233600, 233632, 233695, 233712, 233733, 233781, 233798, 234308, 238716, 238720, 271131, 271328, 271330, 271331, 276129, 276138, 276156, 276162, 276170, 276177, 276183, 276189, 276195, 276218, 276226, 276228, 276239, 276249, 276254, 276255, 276258, 276262, 276264, 276265, 279683, 282927, 282945, 283260, 285876, 285943, 285946, 285961, 285999, 286011, 286018, 286019, 286250, 295694, 295725, 297384, 297473, 297598, 297639, 299635, 299643, 299648, 299662, 299670, 299676, 299689, 300302, 300629, 300634, 300656, 300685, 300690, 300698, 300705, 300709, 300719, 300738, 300744, 300754, 300777, 300780, 300793, 300800, 300806, 300832, 300834, 300850, 300904, 300905, 300917, 300918, 300927, 00931, 300936, 300951, 300955, 300967, 301844, 301857, 301863, 301866, 301867, 301896, 301924, 301942, 301961, 301976, 301981, 302015, 302059, 302102, 302149, 302209, 302224, 302239, 302246, 302341, 304526, 305795, 306179, 307210, 312344, 312358, 312367, 312370, 312371, 312377, 312378, 312382, 312384, 312387, 312391, 312393, 312403, 312415, 312419, 312420, 312423, 312425, 312427, 312435, 312437, 312438, 312443, 312445, 312446, 312451, 313218, 317706, 317710, 317713, 317723, 317726, 317728, 317742, 317744, 317750, 317751, 317761, 317796, 317802, 317817, 317832, 318000, 318012, 318023, 318029, 318042, 318043, 318044, 318045, 318046, 318048, 318054, 318792, 321954, 321955, 321956, 321962, 321972, 321974, 321975, 321977, 321978, 321979, 321988, 321994, 322002, 322015, 322019, 322021, 322028, 322046, 322048, 322065, 322066, 322076, 322087, 322088, 322103, 322112, 322114, 322116, 322123, 322126, 322135, 322140, 322144, 322147, 322149, 322154, 322157, 322162, 322175, 322176, 322178, 322181, 322182, 322183, 322184, 322185, 322186, 322187, 322189, 322190, 322191, 322192, 322194, 322195, 322196, 322197, 322198, 322202, 322203, 322205, 322208, 322209, 322210, 322211, 322212, 322222, 322226, 322231, 322244, 322249, 322259,

322260, 322261, 322269, 322273, 322274, 322279, 322283, 322285, 322311, 323752, 323776, 323803, 323822, 323918, 323968, 324030, 324056, 324137, 324144, 324290, 324855, 324864, 325036, 325048, 325622, 325635, 325653, 325667, 325677, 325700, 325779, 325817, 325850, 325857, 325861, 325867, 325881, 325896, 325908, 325965, 326003, 326047, 326060, 326108, 326116, 326121, 326130, 326139, 326187, 326214, 330307, 330308, 330310, 330311, 330312, 330313, 330314, 330316, 330317, 330318, 330319, 330320, 330321, 330324, 330325, 330327, 330332, 330333, 330336, 330337, 330339, 330341, 330429, 330446, 330465, 330466, 330468, 330476, 330478, 330479, 330480, 330481, 330487, 330491, 330492, 330494, 330496, 330599, 330614, 330616, 330621, 330624, 330625, 330628, 330644, 331269, 331298, 333635, 338420, 338450, 338593, 338598, 338665, 384109 (384150 on tang), 386537, 386502, 386522, 86537, 386632, 386656, 396987, 398554, 398650 (386602 on tang), 398983, 401404, 401374, 21577, 423099.

Marlin Model 1894 or 94 carbine, cal. .44-40, with 15 inch barrel, S/Ns 375792, 386485, 398492, 398586, 407596, 421436, 425082, 427022.

Marlin, Model 1894 carbine, cal. .44-40, with factory fitted 15 inch barrel, S/Ns 386612, 425194.

Marlin Model 1894, caliber .44-40, S/N 376768, with 15.25-inch barrel.

Marlin Model 1895 or 95 carbine, cal. .38-56, with 15-inch barrel, S/N 167406.

Marlin Model 1895 or 95 carbine, cal. .40-65, with 15-inch barrel, S/Ns 313644, 315054.

Marlin Model 1895 or 95 carbine, cal. .45-70, with 15-inch barrel, S/Ns 313669, 313673, 313683, 313684, 313688.

Marlin Model 93, caliber .32 H.P.S., S/N 3375, with 15-inch barrel.

Mauser, Commercial Luger Artillery model semiautomatic pistols cal. 9mm, Mauser banner marked, produced under contract for the Royal Thai Police accompanied by original, German mfd., detachable wooden shoulder stocks.

Mauser, Model 1896 semiautomatic pistol accompanied by original German mfd. detachable wooden holster/shoulder stocks, all semiautomatic German mfd. variations produced prior to 1940, any caliber.

Mauser, Pistol-Carbine, model 1896, 7.63mm, with shoulder stock and 11-3/4" to 16" barrel.

Mauser, Model 1902, 6 and 10 shot magazine capacity, semiautomatic pistols in cal. 7.63 x 25mm (.30 Mauser), having the distinctive hammer safety, barrel lengths of either 3.9 or 5.5", and accompanied by an original detachable wooden holster/shoulder stock.

Mauser, Model 1912/14 pistol, 9mm Mauser short or .45 ACP, with original detachable wooden holster/shoulder stock and 5" barrel.

Mauser Parabellum 75th Year Commemorative Luger Karabiner with accompanying shoulder stock.

E. Mayer single shot handgun, no S/N, cal .38-55 smoothbore, barrel length 14-1/2".

Military type Nambu pistol, model 1904, cal. 8mm Nambu with an accompanying original detachable telescoping wooden holster/shoulder stock.

* Novelty four barrelled palm gun in .22 caliber, manufactured by O.F. Mossberg or C.S. Shattuck Arms Co, circa 1906-1909.

One Pocket Creedmores and other original pocket rifles with extension shoulder stocks, cal. .22, made by Samuel Watson Johnson (1838-1903).

OSS Glove Pistol, caliber .38 S&W or .38 special.

OSS "Liberator" pistol, .45 ACP or 9mm.

Remington Flare (Very) Pistol, Mark III, 10 gauge.

Royal, semiautomatic pistol, cal. 7.63 x 25mm (.30 Mauser) having an integral 10 or 20 round magazine, 5-1/2", 6-1/4", or 7-1/8" barrel, and accompanied by an original Spanish mfd. detachable holster/stock.

Sedgley, Mark V, 10 gauge, signal pistol.

The Shatuck "Unique" palm gun in .22 and .32 cal. rimfire.

Smith & Wesson, model Military and Police revolver, cal. .38, S/N 112037, with original Ideal holster/shoulder stock.

Smith & Wesson, model 40, Light Rifle, caliber 9mm parabellum.

Spanish Star, model A, semiautomatic pistol in cals. 7.63 Mauser, 9mm parabellum, 9mm long, .38 ACP and 45 ACP, accompanied by original Spanish mfd. detachable wooden holster/shoulder stock.

Spanish mfd. copies of the Mauser, model 1896, semiautomatic pistol produced prior to 1946 in cal. 7.63mm or 9mm and having either integral or detachable magazines; accompanied by

original Spanish mfd. detachable wooden holster/shoulder stock.
Stevens, Rifle, No. 20, w/smooth bore barrel for .22 and .32 rimfire shot cartridges.
Stevens, Reliable Pocket Rifle, second issue, cal. .22 long rifle or .22 Stevens-Pope.
Stevens, New model Pocket Rifle, first issue, caliber .22.
Stevens, New model Pocket Rifle, second issue, cal. .22 short, long or long rifle rimfire, .22 WRF, .32 long centerfire.
Stevens, New model Pocket Rifle No. 40, calibers: .22 long, rifle, .22 WRF, .22 Stevens-Pope, and .32 long centerfire.
Stevens, Hunter's Pet No. 34 Pocket Rifle, cal. .22 short rimfire to .44-40 WCF.
Stevens, Vernier Hunter's Pet No. 34 1/2 Pocket Rifle, cal. .22 short .44-40 WCF.
Stevens, Vernier New model Pocket Rifle 40 1/2, cal. .22 long rifle, .22 WRF, .22 Stevens-Pope, .32 long centerfire.
Stevens, 1898 New model Pocket Shotgun, in cals. .38-40 and .44-40.
Stevens, Number 39, New model Pocket Shotgun, cals. .38-40 and .44-40.
Stevens, 44-1/2 rifle, .22 cal. S/N 11006 with two-piece barrel owned by A.O. Neidner, Malden, MA.
Steyr Hahn, model 1911/12, semiautomatic pistol, cal. 9mm Steyr or 9mm parabellum with 5" barrel and accompanied by an original European detachable holsterstock.
Strong, Breech Loading Salute Cannon, 10 gauge, mfd. by Strong Firearms Co., New Haven, CT after 1898.
Swedish, model P-40 Lahti, semiautomatic pistol, in cal. 9mm parabellum, accompanied by original Swedish detachable wooden holster/shoulder stock.
The Taylor "Fur Getter" mfd. by the F.C. Taylor Fur Company, St. Louis, MO, .22 cal. rimfire.
The Taylor "Sure Shot" Trap Sun manufactured by or for the F.C. Taylor Fur Company, St. Louis, Missouri, .38 caliber and having a smooth bore.
U.S. Coast Guard 37mm Deck Cannon, S/N M1-1983.
U.S., Mark II, 10 gauge signal pistols.
Walther, model 1937, pistol, "Armee Pistols," 9mm parabellum, with original detachable shoulder stock and 4.9" barrel.
Webley & Scott pistol, Mark 1, No. 2, .455 cal., with original detachable shoulder stock.
Winchester Model 36, shotgun, 9mm rimfire.
Winchester rifle, Model 1873, .32 caliber, S/N 637480, having a 14-inch barrel.
Winchester Model 1873 carbine, cal. .32-20 WCF, with 14 inch barrel, S/N 695081 B.
Winchester Model 1873 carbine, caliber .32 WCF, with a 15-inch barrel, S/N 382028.
Winchester Model 1873 carbine, cal. .44-40 WCF, with 14 inch barrel, S/Ns 612999B, 710976B, 719510B, 88502.
Winchester Model 1873 carbine, cal. .44-40 WCF, with 15 inch barrel, S/Ns 579080B, ATF-33300-252.
* Winchester Model 1873 carbine, caliber .44 WCF, S/Ns 717912 B and 718977B, with 14-inch barrel.
Winchester Model 1885 carbine, cal. .25-35, with 15-inch barrel S/N 104783.
Winchester Model 1885 carbine, cal. .44-40 WCF, with 15 inch barrel, S/N 102429.
* Winchester Model 1885 rifle, cal. .44 WCF, S/Ns 75808 and 103509, having a 15-inch barrel.
Winchester Model 1885, caliber .44WCF, S/N 103509, with 15-inch barrel.
Winchester Model 1886 Line Throwing Gun, cal. .45-70 smoothbore, with 14-1/2 inch barrel, S/Ns 153830, 154228, 154229, 154231, 154232, 154256, 154257, 154258, 154259, 154260, 154262, 154263, 154264, 154265, 154266, 154267, 154268, 154269, 154270, 154271, 154274, 154287, 154301, 154304, 154305, 154308, 154319, 154320, 154321, 154322, 154327, 154328, 154331, 154332, 154334, 154336, 154337, 164338, 154339, 154340, 154342, 154345, 154346, 154347, 154348, 154349, 154350, 154351, 154352, 154355, 154356, 154357, 154359, 154360, 154365, 154366, 154384, 154385, 154388, 154390, 154393, 154396, 154397, 154398, 154402, 154403, 154404, 154406, 154409, 154411, 154413, 154414, 154415, 154418, 154419, 154420, 154422, 154424, 154427, 154428, 154429, 154437, 154438, 154439, 154440, 154450, 154451, 154452, 154453, 154455, 154463, 154464, 154465, 154466, 154467, 154468, 154471, 154476, 154511, 154512, 154514, 154515, 154516, 154517, 154520, 154521, 154522, 154523, 154524, 154525, 154526, 154527, 154528, 154529, 154531, 154532, 154548, 154550, 154551, 154552, 154554, 154555, 154556, 154557, 154558, 154562, 154564, 154565, 154566, 154567, 154568, 154569, 154570, 154572, 154573, 154574, 154588, 154589, 154590, 154591, 154592, 154595, 154598, 154599, 154600, 154601, 154602, 154604, 154605, 154607, 154608, 154609, 154610, 154611,

154612, 154613, 154616, 154617, 154618, 154619, 154627, 154629, 154631, 154633, 154639, 154640, 154642, 154646, 154648, 154650, 154651, 154653, 154654, 154655, 154660, 154661, 154663, 154664, 154665, 154666, 154669, 154671, 154672, 154676, 154677, 154682, 154688, 154689, 154690, 154693, 154694, 154696, 154697, 154698, 154700, 154702, 154704, 154707, 154708, 154709, 154712, 154713, 154714, 154715, 154718, 154719, 154720, 154721, 154722, 154724, 154725, 154726, 154730, 154731, 154732, 154735, 154737, 154739, 154741, 154742, 154744, 154745, 154747, 154749, 154755, 154764, 154768, 154769, 154771, 154773, 154775, 154776, 154777, 154778, 154779, 154780, 154781, 154782, 154789, 154790, 154791, 154793, 154794, 154795, 154797, 154798, 154799, 154800, 154801, 154802, 154803, 154804,154808, 154814, 154820, 154821, 154822, 154823, 154824, 154825, 154826, 154827, 154830, 154831, 154833, 154834, 154853, 154854, 154855, 154856, 154857, 154858, 154860, 154862, 154865, 154866, 154868, 154869, 154870, 154871, 154875, 154878, 154889, 154891, 154892, 154893, 154894, 154895, 154896, 154897, 154905, 154906, 154912, 154913, 154914, 154928, 154942.

Winchester Model 1892 or 92 carbine, cal. .25-20 WCF with 14-inch barrel, S/Ns 19782, 466396, 575644, 765952, 766291, 800432, *801205, 801232, 801389, 836858, 818777, 859905, 859912, 903891, 940986, 942997.

Winchester Model 1892 or 92 carbine, cal. .25-20 WCF with 15 inch barrel, S/Ns 295206, 512237, 692609, 846245, 852773, 853241, 859948, 982371, 850411.

Winchester Model 1892, caliber .30 WCF, S/N 211284, having a 14-inch barrel.

Winchester Model 1892 or 92 carbine, cal. .32 WCF with 14-inch barrel, S/Ns 353676, 605816, 800188, 807261, 840518, 874184, 928023, 939107.

Winchester Model 1892 or 92 carbine, cal. .32 WCF with 15-inch barrel, S/Ns *475522, 786820.

Winchester Model 1892 or 92 carbine, cal. .32-20 WCF with a 14-inch barrel, S/Ns 615287, 655804, 772706, 807261, 808290, 826022, 837181, 850460, 850556, 859920, 874501, 875174, 960471, 983151.

Winchester, Model 1892, caliber .32 WCF, S/N 841245, with 14-inch octagonal barrel.

Winchester Model 1892 or 92 carbine, cal. .32-20 WCF with a 15-inch barrel, S/Ns 203439, 472150, 475520, 490150, 554549, *690038, 800472, 807957.

Winchester, Model 1892, caliber .35 Winchester S.L., S/N 686515, with 15-inch barrel, marked .25-20 W.C.F.

Winchester Model 1892, cal. .38 WCF, with a 12-inch barrel, S/N 681490.

Winchester Model 1892 carbine, caliber .38 WCF, with a 14-inch barrel, S/Ns *426935, 675600, 686204, 697846, 842774, 848416, 848742, 998336.

Winchester Model 92 carbine, cal. .38 WCF, with a 14-inch barrel, S/Ns 672832, 979585.

Winchester Model 1892 carbine, caliber .38 WCF, with a 15-inch barrel, S/Ns 545168 and 600577.

Winchester Model 1892 or 92 carbine, cal. .38-40 WCF with 12 inch barrel, S/N 290849.

Winchester Model 1892 or 92 carbine, cal. .38-40 WCF with 14 inch barrel, S/Ns 288562, 475736, 653874, 667950, 679709, 847750, 848745, 848873, 850811, 887641, 975236, 998845.

Winchester Model 1892 or 92 carbine, cal. .38-40 WCF with 15 inch barrel, S/Ns 179252, 408307, 449677, 04878, 832039, 842788, 843000.

* Winchester Model 1892, cal. 44-40 WCF, S/N 242340, 999009 with 15-inch barrel.

Winchester Model 1892 carbine, cal. .44 WCF, S/Ns 499886, with factory fitted 14-inch octagonal barrel.

Winchester Model 1892, caliber .44 WCF, S/N 414816, with 14-inch octagonal barrel.

Winchester Model 1892 carbine, cal. .44 WCF, S/Ns *410040, 417473, 518749, with factory fitted 14-inch barrel

Winchester Model 1892 or 92 carbine, cal. .44 WCF, with 14-inch barrel, S/Ns 237383, *241998, 290545, 319311, 323486, 324906, 325125, 325258, 325412, 326732, 353128, 373701, 374202, 394598, 407490, 407735, *409734, 409959, 410015, 410127, 410461, 410488, 411197, 417104, 417274, 471250, 492863, 507295,558135, *567955, 600075, 613228, 659956, 661061, 676111, 688571, 760056, 710257, 742128, 742330, 760298, 761294, 767088, 814474, 845258, 852145, 853128, 861322, 870893, 879896, 880098, 882743, 889143, 894524, 895290, *895626, *896406, 940939,905438, 934212, 940138, *941030, 953433, 953533, 972059, 977206, 978960, 979529, 979915, 98094, 981253, 982178, 982703, 982509, 987936, 996120, ATF-33100012, 89287, 325676, 436742 and 982549.

* Winchester Model 1892, caliber .44 WCF, S/Ns *58732, 168291, 172068, 191413, 302989, 338220,

454616, 449861, 677916, 697497, 756846, 820410, 820507, 820970, 821110,824056, 827272, 835806, 842869, 844314, 850990, 871544, 882743, 889143, 893373, 894230, 895861, 908502, 971295, 985251, 991413, *992152, 995690, 996327, 997543, 998263, 998466, 998995, 998780, 999018, 999036, with 15-inch barrel.

* Winchester Model 1892, caliber .44, S/N 773086, having a 14-inch barrel with an engraved receiver.

Winchester carbine, Model 1892, caliber .44 WCF, barrel length 14 1/8 inches, S/N 895675.

* Winchester rifle, Model 1892, .44 WCF caliber, S/Ns having a 15-inch barrel.

* Winchester Model 1892, caliber .44 WCF, S/N 998929, with factory fitted 15-inch barrel.

Winchester Model 1892, caliber .44 WCF, S/Ns 408070, 693360 with 12-inch barrel.

Winchester Model 1892, caliber 44-40 WCF, S/Ns 172450, *301164, *678255, with 15-inch barrel.

Winchester Model 1892 or 92 carbine, cal. .44-40 WCF, with 14 inch barrel, S/Ns 268828, 319250, 320567, 323075, 323300, 323428, 324318, 325248, 325618, 326147, 340471, 349471, 407898, 409715, 410033, 417996, 446526, 475736, 495045, 583403, 600903, 710267, 743634, 761295, 819003, 820170, 822013, 823943, 824396, 842679, 852636, 861856, 879282, 890745, 895062, 895675, 897168, 898838, 898938, 921643, 925550, 948932, 976596, 977351, 977828, 978462, 979227, 979429, 980055, 980097, 980478, 981212, 982254, 982483, 982562, 984073, 987239, 987305, 989144, 993799, 999398.

Winchester Model 1892 or 92 carbine, cal. .44-40 WCF, with 15-inch barrel, S/Ns 197124, 381573, 414906, 511099, 518748, 650991, 659700, 756846, 816422, 820590, 823559, 837672, 844464, 881573, 886301, 895657, 896494, 905158, 937022, 954796, 977103, 988160, 990107, 991595, 994171, 996523, 997305, 998059, 998419, 998461, 998679, 998741, 998761, 998844, 998852.

Winchester Model 1892, caliber .25-20WCF, S/N 686515, with 15-inch barrel.

Winchester Model 1892 caliber .25-20WCF, S/N 914301, with 15-inch barrel.

Winchester Model 1894 or 94 carbine, cal. .25-35, with 14-inch barrel, S/Ns 427187, 814142, 865268.

Winchester Model 1894 or 94 carbine, cal. .25-35, with 15-inch barrel, S/Ns 155531, 444585, *522192, 542097, 589609, 812956, 815238.

Winchester Model 1894 or 94 carbine, cal. .30-30 WCF, with 14-inch barrel, S/Ns 462548, 505307, 793359, 805815, 842523, 864415.

* Winchester Model 94 lever action rifle, caliber .30 WCF, S/N 990233, with factory fitted 15-inch barrel.

Winchester Rifle, Model 1894, .30 WCF caliber, S/Ns 442920, 463348, 554548, 612946, 639534, 780670, 820363, 844538, 846610, 847112, *865212, 862579, 885127, 885359, 921311, 921754, 921922, *925038, 958505, 992289, 959632, 986535, 1004929, 1004427, 1005373, 1009473, 1017950, 1032419, 1044205, 1050280, 1054200, *1052693, 1056793, 1057984, 1058915, 1065744, 1086186, having a 15-inch barrel.

* Winchester Model 94 carbine, caliber .30 WCF, S/Ns 441082, 846992, 1009372, 1086226, 1009382, 865026, 870711 and 839968. with 15-inch barrel.

Winchester Model 1894 carbine, caliber .30 WCF, S/N 321470 with factory fitted 14-inch octagonal barrel.

* Winchester, Model 94 carbine, caliber .30 WCF, S/Ns 413187, 746707, 809300, 916487 with 14-inch barrel.

Winchester rifle, Model 1894, .30 WCF caliber, S/N 1057890, having a 15-inch barrel.

Winchester Model 1894 carbine, cal. .30 WCF, S/Ns *847116, *910606, 1017717, *1031654, 1033895, with factory fitted 15-inch barrel.

Winchester Model 94 carbine, cal. .30 WCF, S/N 662205, with factory fitted 15-inch barrel.

Winchester Model 1894 or 94 carbine, cal. .30 WCF, with a 15-inch barrel, S/Ns 502433, 503849, 505391, 507520, 639857, 640789, 758048, 801966, 820102, 864818, 885107, 885500, 909096, 910366, 923115, 923639, 937406, 959256, 981707, 992287, 998264, 1017931, 1021206, 1029281, *1033896, 1034950, *1038930, 1040702, *1041205, 1046882, 1040987, 1065446, 1065843, 1066548.

Winchester Model 1894, .30 WCF caliber, S/Ns 1052837, 933906, 910064, 922873 having a 15-inch barrel.

Winchester Model 1894 carbine, caliber .30 WCF, S/N 554734, with factory fitted 15 1/8-inch barrel.

Winchester Model 1894, cal. .30-30 WCF, S/N 848830, with original 15-inch unserviceable barrel.

Winchester Model 1894 or 94 carbine, cal. .30-30 WCF, with 15-inch barrel, S/Ns 273691,315250,

318257, 324318, 360587, 368971, 373033, 433426, 444269, 447421, 464604, 467286, 478529, 481877, 486201, 491869, 495342, 554439, 554727, 559831, 593839, 602691, 638701, 639264, 661329, 662192, 701730, 758406, 790018, 790761, 794467, 802743, 806780, 815570, 820101, 820375, 820413, 839880, 840123, 857252, 862245, 868769, 868912, 869576, 870637, 871116, 871276, 872505, 882967, 883055, 884272, 885127, 885975, 887409, 910758, 923215, 928280, 933642, 933902, 938370, 943237, 952507, 958812, 958947, 959421, 959486, 972583, 973001, 973063, 976472, 981452, 981456, 986621, 990851, 991114, 992557, 995540, 995578, 995675, 998829, 1004615, 1004958, 1009373, 1009414, 1013465, 1025302, 1032798, 1033898, 1033973, 1034037, 1040905, 1046737, 1051727, 1051776, 1052211, 1052749, 1052851, 1052982, 1057043, 1059318, 1066951, 1068292, 1072755, 1080781, 1086226, 1090460

Winchester Model 1894 or 94 carbine, cal. .32 Win. Spl., with 14-inch barrel, S/N 781211, 889043.

Winchester Model 1894 or 94 carbine, cal. .32 Win. Spl., with 15-inch barrel, S/Ns 437440, 840832, 859421, 860098, 880845, 1006715.

Winchester Model 1894 or 94 carbine, cal. .32-40 WCF, with 15-inch barrel, S/N 355640, 894588.

Winchester Model 1894 or 94 carbine, cal. .38-55 WCF, with 15-inch barrel, S/N 247646.

Winchester, 1902, pistol made from a rifle, cal. .22, produced at factory, having no S/N, but having a decorated stock and accompanied by documents denoting its use by Mr. Ned Tupperwein.

Winchester, 1892 Saddle Ring Carbine, .32 WCF cal. with 15" barrel, S/Ns 575423 and 838742.

Winchester, 1892 Saddle Ring Carbine, .38 WCF cal. with 14" barrel, S/N 683395.

Winchester, 1892 Saddle Ring Carbine, .44 WCF cal. with 14" barrel S/N 988293.

Winchester, 1894 Saddle Ring Carbine, .30 WCF cal. with 15" barrel S/N 959111, 1040690 and 986607.

Winchester, Model 1894 Saddle Ring, .30 WCF cal. with 15 1/16 " barrel S/N 933879.

SECTION IIIA: Weapons Removed From The NFA As Collector's Items And Removed From the GCA As Antiques.

The following firearms were removed from the NFA as collector's items and classified as curios or relics under 18 U.S.C. Chapter 44. However, because they are antiques as defined in Chapter 44, they should not have been classified as curios or relics. Because they are no longer NFA weapons and are antiques under Chapter 44, they are not subject to GCA provisions.

Any pistol or revolver, mfd. in or before 1898, originally designed to accept a shoulder stock, and accompanied by an original shoulder stock.

Belgian, Cane Gun, 41 caliber rimfire.

Bergmann, model 1897, cal. 7.65mm (7.8mm) pistol with accompanying shoulder stock.

Borchardt, model 1893, cal. 7.63mm pistol with accompanying shoulder stock.

Chicago, palm pistol, caliber .32 rimfire extra short.

Frank Wesson, Bicycle Rifle with accompanying shoulder stock.

French, Model 1897 75mm breechloading cannon of French manufacture, mfd. in or before 1898.

Gaulois, palm squeezer, 8mm short "Little All Right" palm pistol, .22 cal. rimfire patented by Edward Boardman and Andrew Peavey, January 18, 1876.

German Minenwerfers, old and new models, mfd. prior to 1919, utilizing radial, straight pull friction primer ignition system.

Mannlicher, Pistol-Carbine, model 1896, 7.63mm Mannlicher, with rifle type shoulder stock and forearm and 11-3/4" barrel.

Marlin Model 1889, cal. .32, with 15-inch barrel, S/Ns 45952, 69571, 70624, 70678, 70729, 70737, 70738, 80884, 85429, 96169.

Marlin Model 1889, cal. .38, with 15-inch barrel, S/Ns 53545, 63448, 63476, 63478, 69566, 69568, 69569, 69657, 69660, 73133, 73715, 73742, 83805, 83875, 83892, 83949, 83957, 83984, 83986, 88942.

Marlin Model 1889, cal. .44, with 15-inch barrel, S/Ns 40013, 40021, 40074, 40083, 40084, 40089, 44305, 44306, 44313, 44324, 44327, 44330, 44332, 44335, 44336, 44338, 44343, 44345, 44347, 44348, 44351, 44360, 44366, 44369, 44372, 46621, 46641, 46709, 46743, 46765, 46876, 48997, 49009, 49012, 49016, 49019, 49026, 49029, 49031, 51902, 51912, 52022, 52023, 52029, 52031,52038, 52039, 52043, 53304, 53318, 53360, 53376, 53392, 53393, 55365, 55820, 59140, 59792, 61252, 61265, 61347, 61350, 61903, 61904, 61929, 61932, 61935, 61972, 61982, 61985, 63440, 63458, 63732, 63735, 63747, 63761, 63763, 65070, 65074, 66329, 66415, 66819, 66841, 66842, 66846, 66852, 66888, 67397, 67402, 67403, 67406, 67433, 67436, 67454, 67457, 67460, 67513, 67518, 68599, 68612, 68620, 70779, 70784, 70821, 70830, 70832, 72848, 73049, 73083, 73085, 73090, 73151, 73159, 73160, 73181, 73195, 73208, 73216, 73218, 73713, 73729, 73733, 73735, 73749, 73750, 73751, 73764, 73768, 73772, 73774, 73776, 73789, 73814, 73824, 73838, 73842, 73848, 73853, 81046, 81052, 81063, 81077, 81084, 81092, 81099, 81119, 81187, 81194, 83456, 83471, 83472, 83484, 83493, 83499, 83501, 83513, 83514, 83515, 83520, 83533, 83542, 83543, 83554, 83559, 83632, 83679, 83691, 83859, 83870, 83930, 83933, 83937, 83950, 83953, 83954, 83964, 83969, 86458, 86506, 86674, 86697, 86745, 86746, 86782, 86786, 86795, 88938, 89484, 89485, 89491, 89503, 89523, 89534, 89535, 89541, 89562, 89584, 89587, 89609, 89818, 89823, 89831, 90012, 90013, 90016, 90021, 90037, 90053, 90055, 90072, 90075, 90706, 91389, 91496, 91520, 91657, 91672, 91819, 91842, 91869, 91887, 91889, 91918, 91949, 91962, 91974, 91981, 91983, 91984, 91990, 91995, 92009, 92010, 92011, 92012, 92030, 92044, 92061, 92074, 92078, 92079, 92104, 92106, 92116, 92133, 92137, 92142, 92144, 92154, 92170, 92172, 92178, 92193, 92197, 92198, 92213, 92218, 92224, 99751, 99753, 99758, 99766, 99819, 99820, 99869, 99885, 99967, 99972, 99974, 99992, 99994, 100003, 100009, 100021, 100086, 100228, 108487, 108513, 108517, 108526, 108532, 108537, 108541, 108546, 108554, 108555, 108557, 108567, 108570, 108571, 108572, 108575, 108577, 108579, 108581, 108585.

Marlin Model 1893, cal. .25-36, with 15-inch barrel, S/N 124940, 374465.

Marlin Model 1893, cal. .30, with 15-inch barrel, S/Ns 161903, 162931, 164888.

Marlin Model 1893, cal. .32, with 15-inch barrel, S/Ns 96935, 103641, 103643, 107944, 107948, 107950, 107958, 107970, 124405, 131030, 142050, 147644, 147669, 162785, 162810, 176191.

Marlin Model 1893, cal. .38-55 or .38-56, with 15-inch barrel, S/Ns 96930, 96938, 96942, 103642, 103648, 103660, 103661, 103666, 103670, 133880, 147638, 147658, 167090, 167406.

Marlin Model 1894, cal. .25, with 15-inch barrel, S/Ns 137247, 137248, 137250, 137252, 137256, 137261, 140941, 146754, 146755, 146756, 146757, 146758, 146759, 147678, 147681, 147682, 147688, 147694,

147699, 147700, 147701, 147703, 148157, 148173, 148183, 148217, 151981, 174745.

Marlin Model 1894, cal. .32, with 15-inch barrel, S/Ns 47431, 84993, 126545, 129211, 129221, 129227, 130657, 134113, 134118, 134126, 135406, 135407, 137124, 137125, 137129, 137130, 137131, 137133, 137136, 137137, 137138, 137139, 137145, 137146, 137147, 137149, 137151, 137156, 140876, 140877, 140892, 140899, 140903, 140907, 140909, 140921, 140922, 140924, 140927, 140931, 140934, 140942, 140943, 140945, 142038, 142039, 147206, 147704, 148147, 148153, 148155, 148158, 148178, 148183, 148184, 148194, 148197, 148200, 148201, 148207, 148209, 148212, 148214.

Marlin Model 1894, cal. .38 WCF, with a 15-inch barrel, S/N 150463.

Marlin Model 1894, cal. .38, with 15-inch barrel, S/Ns 109874, 111674, 115770, 115772, 126799, 126858, 126907, 126953, 127093, 127180, 127382, 133190, 133260, 133491, 133645, 134348, 134363, 134364, 134446, 134495, 134526, 134565, 137669, 142371, 142374, 142379, 142389, 145783, 145784, 145787, 145788, 145789, 145790, 145792, 145793, 145794, 145796, 145798, 145799, 145800, 145806, 145813, 147629, 147630, 147631, 147632, 147633, 147634, 150482, 150483, 150486, 150576, 151586, 151594, 151651, 151721, 151726, 151737, 151746, 151748, 151751, 151753, 151761, 151923, 151987, 152004, 152043, 152057, 152066, 152070, 169850,169851, 169852, 169855, 169856.

Marlin Model 1894, cal. .44, with 15 inch barrel, S/Ns 109588, 111398, 111696, 111789, 111794, 111813, 115362, 115363, 115365, 115378, 115383, 115385, 115388, 115392, 115393, 115400, 115403, 115407, 115408, 115412, 115413, 115414, 115418, 115419, 115421, 115433, 115436, 115438, 115439, 115449, 115451, 115453, 115457, 115458, 115459, 115473, 115476, 116487, 115489, 115496, 115499, 115616, 115622, 115644, 115649, 115651, 115662, 115710, 115723, 115743, 115822, 120065, 120094, 120107, 120115, 120118, 120119, 120126, 120133, 120137, 120139, 120145, 120146, 120149, 120161, 120213, 120227, 120228, 120239, 120243, 120270, 120273, 120276, 120292, 120296, 120469, 120496, 120504, 120505, 120523, 120534, 120549, 121549, 121933, 124651, 124662, 124671, 124704, 124725, 124734, 126787, 126843, 126851, 127149, 127251, 127300, 133605, 134090, 134376, 134543, 137495, 137614, 137725, 137734, 137765, 142367, 142368, 142373, 142376, 142377, 142380, 142390, 142402, 144194, 144197, 144198, 144272, 144296, 145687, 145689, 145693, 148154, 150181, 150260, 150369, 150371, 150380, 150384, 150385, 150406, 150420, 150421, 150430, 150462, 150471, 150568, 151047, 151074, 151082, 151086, 151087, 151091, 151092, 151094, 151096, 151097, 151101, 151107, 151133, 151134, 151137, 151140, 151144, 151145, 151151, 151158, 151187, 151193, 151195, 151214, 151227, 151242, 151243, 151619, 151622, 151631, 161632, 151698, 151723, 151724, 151730, 151733, 151735, 151743, 151744, 151756, 151757, 151760, 151764, 151790, 151792, 151797, 151994, 151997, 152020, 152022, 152075, 169854, 169914, 169940, 169971, 169972, 169986, 170002, 170025, 170138.

* Mauser, Model 1896, semiautomatic pistols or carbines having the following combination of features: 1.) Serial number below 1500. 2.) Large ring or cone hammer. 3.) Early safety (safety on with lever in down position). 4.) Early extractor (straight sided at front without semicircular lobes on either side). 5.) Early firing pin retainer (U-shaped insert in back of bolt which retains firing pin).

Mervellieux, squeezer pistol, 6mm and 8mm short.

Peavey, A.J., Knife Gun, .22 short rimfire.

Protector palm gun, .32 rimfire extra short, patented by Jacques Tublaux, Patent No. 732644.

Quackenbrush Bicycle Rifle, with telescopic wire stock, .22 caliber.

Remington, Cane Gun, model 1, .22 rimfire.

Remington, Cane Gun, model 2, .32 rimfire.

Springfield Trapdoor, Model 1884, cal. 45-70, line throwing gun, having a barrel length less than 16 inches, S/N 294879.

Stevens, Old model Pocket Rifle, .22 short or long rimfire.

Stevens, Reliable Pocket Rifle, first issue, cal. .22 short, long or long rifle.

Stevens, New Model Pocket Rifle, first issue, in cal. .32 short or long rimfire.

Stevens, New Model Pocket Rifle, 2nd issue, in cals. .25 Stevens or .32 long rimfire.

Stevens, New Model Pocket Rifle No. 40, in cal. .25 Stevens or .32 long rimfire.

Stevens, New Model Pocket shotgun, in cal. .44-50 Everlasting.

Stevens, Vernier, New Model Pocket Rifle, cal. .22 short, .22 long rifle, .22 WRF, .32 long rimfire.

Stevens, Vernier, New Model Pocket Rifle, No. 40 1/2, in cal. .25 Stevens or .32 long rimfire.

Tribuzio, "Squeezer" invented by Catello Tribuzio of Turin, Italy, cal. 8mm short.

Winchester Model 1873 carbine, cal. .32-20 WCF, with 15-inch barrel, S/N 382027.

Winchester Model 1873 carbine, cal. .38-40 WCF, with 14-inch barrel, S/N 514709.

* Winchester Model 1873, caliber .44 WCF, S/Ns 131592A and 92835A, with 15-inch barrel.

* Winchester Model 1873 carbine, cal. .44-40 WCF, with 15-inch barrel, S/Ns 69450, 92537, 92842A, 127884,

131762A, 145337A, 198040B, 234795B, 235014B, 247094B, 263826B, 336514B, 336549, 380061, 486139B.

Winchester Model 1885 carbine, cal. .38-40 WCF, with 15-inch barrel, S/N 79085.

Winchester Model 1885 carbine, cal. .44-40 WCF, with 15-inch barrel, S/Ns 73604, 73885, 78928, 79256, 83304.

* Winchester Model 1885 single shot carbine, cal. .44 WCF, with factory fitted 15-inch barrel, S/N 83313.

Winchester Model 1885 carbine, caliber .44 WCF, S/Ns 73593, 75744, *75759, with a 15-inch barrel.

Winchester Model 1892 carbine, cal. .32-20 WCF, with 14-inch barrel, S/N 50480.

Winchester Model 1873 Carbine, caliber .44 WCF, having a 15-inch barrel, S/N 249401B.

Winchester Model 1892 carbine, cal. .38-40 WCF, with 15-inch barrel, S/N 43844.

Winchester Model 1892, caliber .38 WCF, S/N 102645, with 15-inch barrel.

Winchester Model 1892 carbine, cal. .44-40 WCF, with 15-inch barrel, S/Ns 51976, 95516, 158145.

Winchester Model 1894 carbine, cal. .30-30 WCF, S/N 42771.

SECTION IV: National Firearms Act Weapons Classified As Curios Or Relics Under 18 U.S.C. Chapter 44

The Bureau has determined that the following NFA weapons are curios or relics as defined in 27 CFR 478.11 because of their dates of manufacture. These NFA weapons, classified as curios or relics, are still subject to all the controls under the NFA. However, licensed collectors may acquire, hold, or dispose of them as curios or relics subject to the provisions of 18 U.S.C. Chapter 44 and 27 CFR Part 478. They are still "firearms" as defined in 18 U.S.C. Chapter 44 and 26 U.S.C. Chapter 53.

Alarm Pistol, S/N IRS-3591, cal. .22 rimfire, marked Patent Pending about 1883, which rings an alarm bell and also fires a cartridge when set to do so.
Armalite, AR 10, 1955, cal. 7.62mm NATO submachinegun, made by the Armalite Div. Of Fairchild Engine & Airplane Co., Costa Mesa, CA, prior to 1961.
Armalite AR-15, .222 Remington or.223 cals., produced by Armalite.
Armalite AR-15, Model 601,.223 cal., mfd. by Colt. Must be marked "Armalite."
Armalite, AR-18, selective fire rifle, cal. .223, prod. by Armalite, Costa Mesa, CA.
Astra, models 901, 902, and 903, selective fire pistols, cal. 7.63mm (.30 Mauser).
Astra, model F, selective fire pistol, caliber 9mm Bergmann.
Azul, selective fire pistol, cal. 7.63mm (.30 Mauser), copy of Mauser model 1896 pistol.
Australian, Austen MK I and MK II Submachineguns, cal. 9mm, mfd. in Australia prior to 1946.
Australian, Owen MK I and MK II Submachineguns, cal. 9mm, mfd. in Australia prior to 1946.
Austrian, Swartzlose machinegun, model 1905 and 1907/1912, cal. 8mm.
Austrian, Steyr-Solothurn model MP-30 and MP-34 submachineguns as originally mfd. by Steyr-Daimier-Puch of Steyr-Austria, prior to 1946.
Auto Ordnance, WWII commemorative Thompson model 1928, caliber .45.
Auto Ordnance, Korean War Commemorative Thompson model 1928, cal. .45.
Auto Ordnance, Thompson Model 1928, .22 caliber machinegun in original configuration as manufactured at the West Hurley, NY, facility between 1980 and 1981.
Benet Mercie, machine rifle, model 1909 (French and American manufacture), caliber 8mm Lebel or .30.
Boys, MK1, caliber .55 , anti-tank rifle.
Bren, light machineguns, any model, any cal., British Commonwealth manufacture.
Brevettata, model F.N.A.- B 1943, submachinegun, cal. 9mm parabellum,with an 8" barrel, S/N 5391.
British, Lanchester (Lancaster) Mk 1 and Mk 1, 9mm, submachineguns mfd. in England during World War II.
British Model 1903, 4.5 inch howitzer.
British, STEN submachineguns Mk. I, Mk. II, Mk. III, Mk. IV and Mk. V, cal. 9mm, original British Commonwealth Military issue only.
Browning automatic rifle (BAR), model 1918A2, mfd. by New England Arms Corporation prior to 1946.
Browning automatic rifle M1918A2, S/N 00229, mfd. by Group Industries.
Canadian, EX1 and EX2 experimental series of the FN FAL in cal. 7.62mm NATO, as mfd. by Fabrique Nationale, Herstal, Belgium, for the Canadian Government.
Cane Gun, cane with horn handle, silver tipped, wood-covered steel, breech loading, smooth bore barrel, cal. .32 centerfire, S/N IRS-5834.
Cane Gun, English walking stick, bamboo covered steel rifled barrel with horn handle, caliber .38 centerfire, S/N IRS-3589.
Cane Gun, cal. .410 centerfire, S/N IRS-3587.
Chinese mfd. copy of the U.S. 60mm mortar mfd. prior to 1946.
Chinese, Type 24 (Maxim) machinegun, cal. 7.92mm, Chinese manufacture.
Chinese, Type 26 light machinegun.
Colt, CMG-2, S/N 012.
Colt, Machinegun, model 1895, any caliber.
Colt, Machinegun, model 1914, any caliber.
Colt, Model MG38 and MG38B, water cooled machinegun any cal. mfd. by Colt.
Colt, Model 1919 auto rifle.
Colt, Model 1921, Model 1921A1, Model 52, and Model 52A, .50 cal., water-cooled machinegun, as originally mfd. by Colt and bearing the original S/N with the letter "C" prefix.
Colt, Monitor model machine rifle, cal. 30-06, commercial version of the BAR, 18" barrel, vertical pistol

grip, Cutts compensator, mfd. by Colt.
Crescent, Certified Shotgun .410 gauge, with 12-1/4" barrel and pistol grip.
Czechoslovakian, model ZB 26 and ZB 30 light machinegun.
Danish HOVEA M1949, 9mm parabellum, produced in Denmark.
Danish, Madsen machineguns, all models produced prior to 1946; all calibers.
Danish, Madsen, model 1946, submachineguns.
Danish, Madsen, model 1950, submachineguns.
Degtyarev RPD Light Machine Gun, 7.62 x 39mm caliber, mfd. in the Soviet Union, the People's Republic of China, or the People's Republic of Korea (North Korea).
Degtyarev-PTRD 1941, 14.5mm anti-tank rifle.
Egyptian "Port Said" version of the Swedish Model 45B, 9mm cal., produced in Egypt.
Enfield, military bolt action rifle cut-a-ways, having barrels of less than 16".
Fabrique Nationale, model BAR, full automatic rifle, all cals., as mfd. by Fabrique Nationale, Herstal, Belgium.
Federal Laboratories, 12 Gauge Gas Billy, all variations.
French, 25mm anti-tank cannon SA-L mil. 1937, commonly known as French models 34 SA and 37A.
French, Chatellerault model 1924/29 machineguns, cal. 7.5mm.
French, C.S.R.G. (Chauchat) machinegun model 1915, cal. 8mm Lebel.
French, C.S.R.G. (Chauchat) machinegun, model 1918, cal. .30.
French, model MAS 1938, sugmachinegun, caliber 7.65mm Long.
French, model MAT 49, submachinegun, caliber 9mm.
French, model 1897 75mm breechloading cannon of French and American manufacture, in 1899 or later.
French, model 37, 5Omm (1.97 in.) mortar - also designated as a German, model 203(f), 5cm (1.97 in.) mortar.
French, model 34 SA and 37A, antitank cannon 1937 in 25mm.
French, model 1934 M39 (Mle 34 K39), cal. 7.5 x 54mm, aircraft machinegun.
Franchi, model 1962 Semiautomatic carbines, caliber 9mm.
French, St. Etienne machinegun, model 1907, caliber 8mm. Lebel.
Finnish, Tampelia (original), model 1938, caliber 81mm mortars.
German, assault rifles selective fire, produced during 1941 through 1945, in caliber 7.92mm Kurz (7.92 x 33mm).
German, Bergmann model 1915 water cooled machinegun, caliber 7.92mm.
German, Bergmann model 1915, an air cooled machinegun, caliber 7.92mm.
German, Bergmann, MP-34, 9mm, machine pistol.
German, Bergmann, MP-35/1, 9mm, machine pistol.
German, BUCO hand firing device, caliber 12mm (.44 cal.).
German, Dreyse model 1918 water cooled machinegun, cal. 7.92mm.
German, Erma (EMP), or MPE, 9mm, machine pistol.
German Flak 30, 2cm anti aircraft machinegun.
German, Gast machinegun, caliber 7.9mm.
German, FG42 Parachutists rifle, cal. 7.92mm, mfd. prior to 1946.
German, MG 15 anti aircraft machinegun 7.92mm made before 1945.
German, MG 15 machineguns, cal. 7.92mm, made prior to 1945.
German mfd., Model MP3008 submachinegun, cal. 9mm Luger, produced in Germany c. 1944-1945.
German MP41 submachineguns, cal. 9mm parabellum, mfd. by C.G. Haenel, Suhl, Germany during WWII. (Original WWII German production only.)
German Minenwerfers, having fixed firing pins, spring or hammer actuated striker, any cal., any model, mfd. prior to 1919.
German, 15cm Nebelwerfer, 41 (150-mm) Rocket Launcher.
German, Parabellum machinegun, model 1913, 1914 and LMG 14/17, caliber 7.9mm.
German, Raketenpanzerbuchse 43 (8.8cm, RPzB43), model 43, 88mm Rocket Launcher.
German, Raketenpanzerbuchse 54 (8.8cm, RPzB54), model 54, 88mm Rocket Launcher.
German, Schmeisser, MP-28/II, 9mm machine pistol.
German, Steyr-Solothurn, 9mm machine pistol.
German, MG 34 light machinegun, cal. 7.92mm.
German, MG 42 light machinegun, produced prior to 1946, cal. 7.92mm.
German, MG 131 machineguns, all models, all calibers, mfd. before 1946.
German, MP 18-1, submachinegun, cal. 9mm parabellum.

SECTION IV

German, MP 38 submachinegun, cal. 9mm parabellum, mfd. by B. Geipel (Erma) prior to 1941.
German, MP 40 submachineguns, cal. 9mm, mfd. from 1940-1945 for the German military forces.
German, 7.5cm Pak 40 anti-tank gun (75mm).
German, 7.5cm Pak 39 anti-tank gun (75mm).
German, 7.5cm Stuk assault gun (75mm).
German, 7.5cm Kw.K 40 tank gun (75mm).
German, 8mm MG 81 and MG 81Z aircraft Machineguns.
German, 15/20mm MG 151 Aircraft Machineguns (all variations).
German, 30mm MG 108 Aircraft Machineguns/Cannons.
German, 13mm Rheinetal-Borsig model 131, Aircraft Machineguns (all variations).
German 37 mm Rheinmetall Borsig anti tank guns mfd. prior to 1946, including the following variations or designations: German issue 3.7cm PAK 35, PAK 35/36, PAK 36, Swedish 3.7cm PvKan m/38, Italian 37mm Cannone contracarro da 37/45 [captured designation 3.7cm PAK 162(I)], Russian 37mm Protivotankovaya Pushka obrets 1930 [captured desgination 3.7 PAK 158(r)], Dutch 37mm Rheinmetall [captured designation 3.7 cm PAK 153(h)], and Japanese 37mm Type 97 (PAK36 purchassed by Japan).
H&A A. Company, 16 gauge, set gun.
H&R Reising Model 55, submachineguns, caliber .45 ACP, manufactured by Harrington and Richardson, Worcester, MA, prior to 1946.
Harrington and Richardson (H&R) Handy Guns, smoothbore pistols, mfd. at Worcester, MA, all calibers, all barrel lengths, having an overall length of less than 26 inches.
Harrington and Richardson (H&R) Handy guns, with shoulder stock, mfd. at Worcester, MA, having a smoothbore barrel less than 18 inches in length, or having a rifled barrel less than 16 inches in length.
High Standard, USA model HD, .22 1r cal. pistols, originally equipped with silencers for issue to the OSS and other military agencies, S/N range 109110-153890.
Hotchkiss Field Cannon, Model 1909, 37mm caliber, S/N IRS 3857.
Hotchkiss, light machinegun, British model 1909 Mark I and Mark I, caliber .303.
Hotchkiss, models 1897, 1900 and 1914 machineguns of French mfd. in all calibers.
Ingram Models 6 and 7, various cals., produced by Police Ordnance Company.
Ingram Models 8 and 9, various cals., produced in Thailand.
Italian, Breda model 1930 machineguns, all calibers.
Italian, Breda Safat, model 1935, 7mm aircraft machinegun, mfd. by Societa Anonima Ernesto Breda, Brescia, Italy, prior to 1946.
Italian, Breda model 1937 heavy machinegun, caliber 8mm.
Italian, Cannone anticarro e d'accompagnamento 47/32 M1935, cal. 47mm, all variations and manufacturers.
Italian, Fiat Revelli model 1935 machinegun, caliber 8mm.
Italian, Model TZ-45, cal. 9mm submachineguns mfd. before 1946.
Italian, model 38A, 9mm parabellum submachinegun, mfd. by Pietro Beretta SpA, Gardone-val Trompia, Italy, prior to 1946.
Italian, model 38/42, 9mm parabellum, by Beretta SpA, Gardone-val Trompia, Italy prior to 1946.
Italian, Revelli model 1914 machinegun, caliber 6.5mm.
Italian, Scotti machineguns, all models, all calibers, mfd. Before 1946.
Italian, Villar Perosa submachinegun, model 1915, 1916 and 1917, cal. 9mm parabellum.
Ithaca, "Auto and Burglar Guns," mfd. by Ithaca Gun Company from 1922, to 1933, all original guns.
Ithaca Auto & Burglar Gun, 28 gauge, S/N 354442, mfd. in 1921 or 1922.
Japanese, (model 1921 Browning Type), aircraft fixed machinegun, cal. 12.7mm.
Japanese, experimental submachinegun "Bullpup" caliber 8mm Nambu.
Japanese, model 10 and 89, 50 mm grenade dischargers.
Japanese, model 38 Field Cannon.
Japanese, model 94, 37mm gun mfd. prior to 1946.
Japanese, model 94, 37mm gun, mfd. prior to 1946.
Japanese, modified Bergmann (SIG), submachinegun, cal. 7.63mm, featuring bayonet mounting lug.
Japanese, M1932 76mm Battalion Pack Howitzer.
Japanese, Type 1 (1941), aircraft flexible machinegun, cal. 7.92mm, twin barrel type.
Japanese, Type 3 (1927), heavy machinegun, cal. 6.5mm. twin barrel type.
Japanese, Type 11 (1922) light machinegun, all variations, caliber 6.5mm.
Japanese, Type 38 (105) machinegun (Hotchkiss) cal. 6.5mm.

SECTION IV

Japanese, Type 89 (1929) Vickers pattern machineguns, cal. 7.7mm.
Japanese, Type 91 (1931), tank machineguns (modified Type 11), cal. 6.5mm.
Japanese, Type 92 (1932), heavy machinegun, Hotchkiss pattern cal. 7.7mm.
Japanese, Type 92 (1932), light machinegun or aircraft machinegun, Lewis pattern, cal. 7.7mm.
Japanese, Type 92 (1932), tank machinegun, caliber 13.2mm.
Japanese, Type 93 (1938), aircraft machinegun cal. 7.29mm.
Japanese, Type 94, 37mm gun, mfd. prior to 1946.
Japanese, Type 96 (1936), light machinegun cal. 6.5mm.
Japanese, Type 97 (1937), tank machinegun, cal. 7.7mm.
Japanese, Type 97, 20mm Anti-Tank machinegun.
Japanese, Type 98, 7.92 aircraft machinegun, Japanese mfd. prior to1946.
Japanese, Type 99 (1939), light machinegun, all varieties, cal. 7.7mm.
Japanese, Type 100 (1940), aircraft flexible machinegun, cal. 7.92mm, twin barrel type.
Japanese, Type 100 submachinegun, caliber 8mm Nambu, all varieties.
Japanese, 12.7 (13) mm Browning HO-103 machineguns.
Japanese, 20mm Browning Aircraft HO-5 machineguns/cannons.
Japanese, 20mm Oerlikon model 99, machineguns/cannons.
Japanese, 30mm Browning Aircraft HO-155 Type 1, machineguns/cannons.
Japanese, 30mm Browning Aircraft HO-155 Type 2, machineguns/cannons.
Japanese, 37mm Browning Aircraft HO-204 machineguns/cannons.
Japanese, 40mm Caseless Aircraft machineguns/cannons.
Lahti, model 39, caliber 20mm anti-tank rifle.
Lewis, light machinegun (American or British manufacture), caliber .303 or .30.
Lovelace Fur Getter; patented by C.D. Lovelace, June 2, 1914, mfd. by Hopkins and Allen Company, Norwich, CT, 12 gauge.
Luger, Mauser commercial manufacture, semiautomatic pistol, 75 Jahre, Parabellum-Pistol, Mod. Karabiner, Commemorative, caliber 9mm, accompanied by a carbine type shoulder stock.
Madsen light machineguns, Series A, all calibers, mfd. in Denmark between 1936 and 1956.
Marble's Game Getter, Firearms, with a combination of .44 cal. or .410 gauge smooth bore and .22 cal. barrels, both lst and 2nd models, with barrel lengths of less than 18", mfd. by the Marble Arms Corporation of Gladstone, MI, prior to 1943.
Marlin, aircraft machinegun, model 1917 and 1918, caliber .30.
Marlin, U.S. model 1917, ground machinegun, caliber 30/06.
Marlin, tank machineguns, model 1918 caliber .30.
Mauser, mfd. Schnellfeuer selective fire pistols and all original prototypes, cal. 7.63mm Mauser.
Maxim, machineguns of German manufacture, all models and cals.
Maxim, machineguns produced by Maxim Nordenfeldt and Vickers Sons and Maxim from approximately 1885 through 1908, all calibers.
Maxim, machineguns, Russian model 1905, Russian manufacture having a brass water jacket in caliber 7.62mm, Russian rimmed.
Oerlikon, MG-FF and MG-FFM, 20mm aircraft cannon.
Oerlikon, 20mm Automatic Cannon, all variations mfd. in the United States prior to 1946.
* Phoenix Cartridge Company, Model PCMP submachinegun, caliber 9mm or .45, S/N 2.
Pipe Pistol, S/N 420, cal. .22 rimfire (word "pipe" means smoking tobacco type).
Pipe Pistol, S/N IRS-3579, cal. 7mm centerfire, marked drgm Bucksom (word "pipe" means smoking tobacco type).
Reinmetall MG17, 7.9mm aircraft machinegun.
Reising, submachineguns U.S. model M-50, .45 cal., Harrington & Richardson Arms, Worchester, MA, USA, Pat Pending.
Rexer, original British made Madsen type light machineguns mfd. at the British Arms Factory known as Rexer.
Royal, selective fire pistol, cal. 7.63mm (.30 Mauser), copy of Mauser model 1896 pistol.
Russian, Degtyarev machineguns, models DA, DP, DPM, DT, and RP46, caliber 7.62mm.
Russian, model 1910 (Maxim), machinegun, any cal., Russian manufacture.
Russian, model PPD 1943/38 and PPD 1940, submachineguns, cal. 7.62mm, original Russian manufacture.
Russian, 50mm Mortar Launcher.
Russian, PSH41, 7.62mm submachineguns and all copies/variations produced in the Communist Bloc countries prior to mid 1950's.

Russian, PPS42, submachineguns, caliber 7.62mm.
Russian, PPS43 submachineguns, cal. 7.62mm mfd. in the Soviet Union prior to 1946.
Russian, model SG43 heavy machinegun, cal. 7.62mm, Russian manufacture.Russian (U.S.S.R.), Federov, automatic rifle, model 1916 AVF (Avtomat), cal. 6.5 x 5O.5SR (6.5mm Japanese), of Russian manufacture.
Russian (U.S.S.R.), Simonov, automatic rifle, model 1936 AVS (AVS36), cal. 7.62 x 64R, of Soviet manufacture.
Russian (U.S.S.R.), Tokarev, automatic rifle, model 1940 AVT (AVT40), cal. 7.62 x 54R, of Soviet manufacture.
Russian (U.S.S.R.), Stechkin, automatic pistol, model APS, cal. 9 x 18 (Makarov), of Soviet manufacture.
Sedgley, Glove Pistol .38 centerfire, S/N IRS-3580, single shot, marked MK2S, mfd. by Sedgley, Philadelphia, for the U.S. Marines.
Simonov-PTRS 1941, 14.5mm anti-tank rifle.
Smith & Wesson, Model 76 submachineguns, 9mm parabellum caliber, all S/Ns, originally manufactured by Smith & Wesson prior to 1975.
Smith & Wesson, Model 76 submachinegun, cal. 9mm parabellum, S/N range from U100 to U6100.
Solothurn Maschinengewehr, Model 38 (MG38), 7.92mm machinegun, mfd. between 1930 - 1935, by Waffenfabrik Solothur AG, Solothurn, Switzerland.
Solothurn, models S18 100, S18 1000 and S18 1100 20mm semiautomatic, antitank rifle.
Springer, .32 cal., Knife-Pistols, marked "C-16" and "C-12" on the left and right extractors, respectively, with "BAZAAR" marked on the left side of the blade. Some devices are marked "Solingen" on the left side of the blade and "Made in Germany" on the other.
* Springfield, .50 caliber M3 aircraft basic machinegun, S/N 1254946.
SS1-Sidewinder Machinegun, S/N Ex-016.
STAR, models AD, BD, MD, and PD, Selective Fire Pistols, any cal., produced prior to 1953.
Stevens, Offhand Number 35, .410 gauge shot pistols.
Stevens, Auto Shot Number 35, .410 gauge shot pistols.
Stoner, model 63 and 63A, machineguns, all variations, as originally manufactured from 1963 through 1970.
Suomi, model 26 and 31 submachinegun.
Swedish, model 37 and 39 submachinegun.
Swedish, Model 45, and Model 45B 9mm parabellum cal., produced in Sweden.
Swedish, model M-42, 20mm recoilless anti-tank rifles.
Swedish, 3.7mm Bofors anti-tank guns produced prior to 1946, including the following variations and designations: SwedishBofors, 37mm PvKan 34, Polish 37mm Armata przeciwpancerna wz. 36 (German captured designation 3.7 cm PAK 36(p), Danish 37mm antitank gun, designation unknown German captured designation 3.7 cm PAK 157(d), and British Ordnance QF 37mm Mk.1.
Swedish, model 21, 6.5mm, auto rifles mfd. prior to 1946.
Swedish, model 37, 6.5mm, auto rifles mfd. prior to 1946.
Swiss (Societe Industrielle Suisse) NKPO carbines, 7.65mm, with barrels less than 16".
Thompson, model 1927, semiautomatic carbine, caliber .45 ACP.
Thompson, submachineguns including all models, prototypes, & variations actually mfd. in or before 1945.
Thompson submachinegun, Model M1, .45 cal., mfd. by Auto Ordnance Corporation, West Hurley, New York, between 1985 and 1986, having S/Ns from M100A to M717A.
Thompson, model 1928, Korean War Commemorative submachineguns, cal. .45 produced by Auto Ordnance, West Hurley, NY.
Tippman Arms Company, .22 cal. machinegun set, consisting of a Model 1919-A4, S/N 333, Model 1917, S/N 333-A, Model .50 HB, S/N 333-B.
Tippman Arms Company ½ scale Browning Model 1917, water cooled, cal. .22 machineguns.
Tippmann ½ scale, Browning Model 1919 and .50HB machineguns, caliber .22.
Trejo, selective fire machineguns, caliber .22, mfd. in Mexico, circa 1952 to 1972.
Umbrella Gun, 6-shot pepper box revolver with dagger attachment extending from muzzle, combined with handle and umbrella, S/N 16, cal. .22 rimfire, with markings Brevette PT, mfd. by Marquis de Fabrique, France.
United Defense Supply Corporation, model U.D. 42, submachinegun, cal. 9mm parabellum, as mfd. by Marlin
Firearms Company, New Haven, CT.
* United States Field Gun, Model 1902, 3-inch caliber, produced prior to 1916.

SECTION IV

U.S. Arms Company, knife pistol, cal. .22, mfd. prior to 1934.

U.S. Browning machineguns (original), cal. .30, M1917 and M1917A1, mfd. by Colt's Patent Firearms Company, New England Westinghouse Company, Remington Arms, Union Metallic Cartridge Company, Rock Island Arsenal, High Standard Company, Savage Arms Corporation, Buffalo Arms Corporation, Frigidaire, AC Spark Plug, Brown-LipeChappin, Saginaw Steering Gear Division, and Kelsey Hayes Wheel Company.

U.S. Browning 1918 and 1918A2 BAR, mfd. by Colt, Winchester, Marlin Rockwell, New England Small Arms, IBM, and Royal Typewriter.

U.S. Browning model 1919, cal. .30, machineguns, produced by various U.S. Government contractors prior to 1946, all variations.

U.S. Browning model 1919A4 air cooled machinegun, produced by various U.S. Government contractors prior to 1946, cal. .30.

U.S. Carbine, cal. 30, M2, original U.S. Military issue, mfd. prior to 1946, and marked "M-2" on the receiver.

U.S. Johnson models 1941 and 1944 light machineguns, cal. .30.

U.S. M2 Browning, .30 cal. machineguns (ANM2), as mfd. by various U.S. Government contractors prior to and during World War II.

U.S. M 4 Survival Rifle, cal. .22 Hornet, as issued by the government, mfd. in the United States and Canada by Harrington & Richardson Arms Company.

U.S. Machinegun, Model T161, caliber 7.62, S/Ns EXP21, EXP25, EXP26, EXP27, EXP30, EXP32, EXP34, EXP35, EXP37, EXP38, EXP42, EXP45, EXP47, EXP47, EXP51, EXP53, EXP56, EXP57, EXP60.

U.S. Maxim machinegun, model 1904, caliber .30.

U.S. Model M3 and M3A1 submachinegun, cal. .45, original Guide Lamp or Ithaca Gun Co. manufacture only, mfd. prior to 1954.

U.S. M3 submachinegun, 9mm parabellum, as originally produced for the U.S. Government, having no manufacturer's name or S/N and originally marked U.S. 9mm SMG on the left side of the magazine housing, original production only.

U.S. Model M6, 37mm, gun, produced prior to 1946.

U.S. Model T-48 assault rifles, caliber 7.62mm, mfd. by H&R.

U.S. Models Ml, M9 and M18, cal. 2.36" Rocket launcher (Bazooka) and variations thereof.

U.S. Models 1915 and 1918 Vickers machineguns mfd. by Colt in cals. 30/06 or 11mm.

U.S. Mortar (original), cal. 60mm, M1 and M2, with mount M2.

U.S. M6 Survival rifle/shotgun, cal. .22/410 gauge, having 14-1/2" barrels, produced by various U.S. Government contractors.

* U.S. Pack Howitzer, Model M1-A1, 75mm caliber, S/N 336, manufactured in 1941.

U.S. Small Arms Company knife pistol, cal. .22 original manufacture produced prior to 1934.

U.S. 3" Anti-tank Gun, model M5.

U.S. 37mm Antitank guns M3 and M3A1, produced by various U.S. Government contractors prior to 1946.

U.S. 76mm Tank Cannon, Models M1, M1A1, M1A2.Vickers commercial water cooled machinegun, any configuration, any cal., British manufacture.

Vickers original military machineguns, all variations, all cals., mfd. prior to 1946.

Winchester model 92 Trapper carbines having barrel lengths of less than 16" (original Winchester manufacture only).

Frequently Asked Questions

Licensed Collectors

What modifications can be made on C&R firearms without changing their C&R classification?

The definition for curio or relic ("C & R") firearms found in 27 C.F.R. § 478.11 does not specifically state that a firearm must be in its original condition to be classified as a C&R firearm. However, ATF Ruling 85-10, which discusses the importation of military C&R firearms, notes that they must be in original configuration and adds that a *receiver* is not a C&R item. Combining this ruling and the definition of C&R firearms, the Firearms Technology Branch (FTB) has concluded that a firearm must be in its original condition to be considered a C&R weapon.

It is also the opinion of FTB, however, that a minor change such as the addition of scope mounts, non-original sights, or sling swivels would not remove a firearm from its original condition. Moreover, we have determined that replacing particular firearms parts with new parts that are made to the original design would also be acceptable—for example, replacing a cracked M1 Grand stock with a new wooden stock of the same design, but replacing the original firearm stock with a plastic stock would change its classification as a C&R item.

What is a firearm curio or relic?

Firearm curios or relics include firearms which have special value to collectors because they possess some qualities not ordinarily associated with firearms intended for sporting use or as offensive or defensive weapons. To be recognized as curios or relics, firearms must fall within one of the following categories:

1) Have been manufactured at least 50 years prior to the current date, but not including replicas thereof; or
2) Be certified by the curator of a municipal, State, or Federal museum which exhibits firearms to be curios or relics of museum interest; or
3) Derive a substantial part of their monetary value from the fact that they are novel, rare, bizarre, or from the fact of their association with some historical figure, period, or event.

[27 CFR 478.11]

Is there a specific license that permits a collector to acquire firearms in interstate commerce?

Yes. The person may obtain a collector's license; however, this license applies only to transactions in curio or relic firearms. The principal advantage of a collector's license is that a licensed collector can acquire curios or relics in interstate commerce.

[27 CFR 478.41(c), (d), 478.50(b) and 478.93]

How does one get a collector's license?

Submit ATF Form 7CR, Application for License (Collector of Curios or Relics), with the appropriate fee in accordance with the instructions on the form. These forms may be obtained from the Firearms and Explosives Licensing Center in Atlanta, Georgia, your local ATF office, or downloaded from ATF's Internet site (www.atf.gov).

[27 CFR 478.41(c)]

Does a collector's license afford any privileges to the licensee with respect to acquiring or disposing of firearms other than curios or relics in interstate or foreign commerce?

No. A licensed collector has the same status under the Gun Control Act (GCA) as a nonlicensee except for transactions in curio or relic firearms.

[27 CFR 478.93]

Does the Federal firearms law require licensed collectors to comply with State laws and local published

ordinances which are relevant to the enforcement of the GCA?

Yes. It is unlawful for any licensed collector to sell or deliver any firearm or ammunition to any person if the person's purchase or possession would be in violation of any State law or local published ordinance applicable at the place of sale or delivery.

[18 U.S.C. 922(b)(2), 27 C.F.R. 478.99(b)(2)]

Does a license as a collector of curio or relic firearms authorize the collector to engage in the business of dealing in curios or relics?

No. A collector's license only enables the collector to transport, ship, receive, and acquire curios and relics in interstate or foreign commerce, and to make disposition of curios and relics in interstate or foreign commerce, to any other licensee, for the period stated on the license. A collector's license does not authorize the collector to engage in a business required to be licensed under the Act. Therefore, if the acquisitions and dispositions of curios and relics by a collector brings the collector within the definition of a manufacturer, importer, or dealer, he shall qualify as such. A dealer's license must be obtained to engage in the business of dealing in any firearms, including curios or relics.

[18 U.S.C. 922(a) and 923(a)(1), 27 CFR 478.41(c)(d)]

What does "engaged in the business" mean?

The term "engaged in the business," as applicable to a firearms dealer, is defined as a person who devotes time, attention, and labor to dealing in firearms as a regular course of trade or business with the principal objective of livelihood and profit through the repetitive purchase and resale of firearms, but such term shall not include a person who makes occasional sales, exchanges, or purchases of firearms for the enhancement of a personal collection or for a hobby, or who sells all or part of his personal collection of firearms.

[27 CFR 478.11]

Are licensed collectors transfers of curio or relic firearms subject to the Brady law, including the provision for making background checks on transferees?

No. Transfers of curio or relic firearms by licensed collectors are not subject to the requirements of the Brady law. It is, however, unlawful to transfer a firearm to any person knowing or having reasonable cause to believe that such person is a felon or is within any other category of persons prohibited from receiving or possessing firearms.

[18 U.S.C. 922 (d), 27 CFR 478.32 (d), 18 U.S.C. 922(t)]

Is the transfer of a firearm by a licensed dealer to a licensed collector subject to the Brady law?

The Brady law does not apply to the transfer of a curio or relic firearm to a licensed collector, but a licensed collector who acquires a firearm other than a curio or relic from a licensee would be treated like a nonlicensee, and the transfer would be subject to Brady requirements.

Are licensed collectors required to execute ATF Form 4473 for transactions in curio or relic firearms?

No. Licensed collectors are only required to keep a "bound book" record.

[27 CFR 478.125(f)]

Are licensed collectors required to comply with the requirements that written notification be given to handgun transferees and signs be posted on juvenile handgun possession?

The requirement that written notification concerning juvenile handgun possession be given by licensees to a nonlicensee to whom a handgun is delivered applies to curio or relic handguns transferred by licensed collectors. Licensed collectors may lawfully dispose of curio or relic handguns away from their licensed premises. Therefore, the signposting requirement does not apply to licensed collectors since, in the case of collectors, a requirement to post signs at the licensed premises would serve no purpose because the premises is not a business premises open to the public.

[18 U.S.C. 922 (x), 27 CFR 478.103]

Can a licensed collector sell a curio or relic shotgun or rifle to a nonlicensed resident of another State?

Yes. A licensed collector is specifically authorized to sell a curio or relic shotgun or rifle to a nonlicensed resident of another State so long as 1) The purchaser meets with the licensee in person at the licensee's premises to accomplish the transfer, sale, and delivery of the rifle or shotgun; and 2) The sale, delivery, and receipt of the rifle or shotgun fully comply with the legal conditions of sale in both such States.

[27 CFR 478.96 (c)(1)]

Are licensed collectors required to turn in their acquisition/disposition records to ATF if their collector's license is not renewed or they discontinue their collecting activity?

No. The GCA requires the delivery of required records to the Government within 30 days after a firearm "business" is discontinued. A license as a collector of curios or relics does not authorize any business with respect to firearms. This is in contrast to firearm importers, manufacturers, and dealers who are licensed to engage in a firearms business. Therefore, the records required to be kept by licensed collectors under the law and regulations are not business records and are not required to be turned in to ATF when collectors' licenses are not renewed or collecting activity under such licenses is discontinued.

[18 U.S.C. 923(g)(4), 27 CFR 478.127]

May a licensed collector obtain NFA firearms in interstate commerce?

Only if the firearms are classified as curios or relics, are registered in the National Firearms Registration and Transfer Record, and are transferred in accordance with the provisions of the NFA.

[27 CFR 479.83 – 479.86]

May semiautomatic assault weapons, which have been classified as curios or relics, be imported?

Not unless they are being imported for sale to a government agency or law enforcement officer employed by such agency for official use. Because ATF will not approve an importation that would place the importer in violation of the law, ATF would not authorize the importation of semiautomatic assault weapons, even if classified as curios or relics, unless the importer provided evidence that the weapons were being imported for sale to a governmental entity or other exempt purchaser.

[18 U.S.C. 922(v)]

Made in the USA
Monee, IL
04 November 2026

67375801R00036